EXPLORERS'
BOTANICAL NOTEBOOK

EXPLORERS' BOTANICAL NOTEBOOK

In the Footsteps of Theophrastus, Marco Polo, Linnaeus, Flinders, Darwin, Speke and Hooker

Text Florence Thinard   **Photographs** Yannick Fourie

FIREFLY BOOKS

A FIREFLY BOOK

Published by Firefly Books Ltd. 2016

First printing

Publisher Cataloging-in-Publication Data (U.S.)

Names: Thinard, Florence, 1962–, author. | Fourie, Yannick, photographer. |
Collishaw, Barb, translator,
Title: Explorers' botanical notebook : in the footsteps of Theophrastus, Marco
    Polo, Linnaeus, Flinders, Darwin, Speke and Hooker / author, Florence
    Thinard; photographer, Yannick Fourie ; translated by Barb Collishaw.
Description: Richmond Hill, Ontario, Canada : Firefly Books, 2016. | Includes
    bibliography and index | Previously published by Plume de Carotte, France,
    as L'herbier des explorateurs | Summary: "This book follows the journey of
    over 80 pioneering botanists and the important findings and collections they
    have made. It includes each journey and routes taken with the help of maps
    and personal notes" – Provided by publisher.
Identifiers: ISBN 978-1-77085-763-6 (hardcover)
Subjects: Botanists -- Biography.
Classification: LCC QK26.T556 |DDC 580.922 – dc23

Library and Archives Canada Cataloguing in Publication

Thinard, Florence, 1962–
[Herbier des explorateurs]
    Explorers' botanical notebook : in the footsteps of Theophrastus, Marco
Polo, Linnaeus, Flinders, Darwin, Speke and Hooker / Florence Thinard.
Translation of: L'herbier des explorateurs : sur les traces de Thoophraste,
Jussieu, La Porouse, Darwin, Livingstone, Monod.
Includes bibliographical references and index.
ISBN 978-1-77085-763-6 (hardback)
    1. Botanists--Travel. 2. Botanists--Biography. 3. Botany--History. 4.
Scientific expeditions--History. I. Title. II. Title: Thinard, Florence, 1962 .
Herbier des explorateurs
QK26.T45 2016   580.92'2   C2016-902982-4

Published in the United States by   Published in Canada by
Firefly Books (U.S.) Inc.   Firefly Books Ltd.
P.O. Box 1338, Ellicott Station   50 Staples Avenue, Unit 1
Buffalo, New York 14205   Richmond Hill, Ontario L4B 0A7

Translation: Barb Collishaw

Printed in China

This book was created in partnership with the Royal
Botanic Gardens, Kew, and the Institut de botanique
de Montpellier, however the publisher retains sole
responsibility for the textual content.

The herbarium plates appearing on the right-hand
pages come from the Herbarium at Kew except the
plates on pages 49, 69, 75, 83, 93, 95, 101, 155 and
161, which are from the herbarium collections of the
Institut de botanique de Montpellier (Université 2).

Reproduction authorized by the Board of Trustees of
the Royal Botanic Gardens, Kew and the administra-
tors of the Université 2 Montpellier.

Many of the historic documents reproduced here are
held by the Library, Art and Archives at Kew.

For Plume de carotte
Text: Florence Thinard; Photographs of herbarium
pages: Yannick Fourié; Artistic director: Geneviève
Démereau; Layout: Cédric Cailhol; Managing editor:
Laura Puechberty; Copy editor: Claire Debout

CONTENTS

Dedicated to my mother, for her love of flowers, trees and gardens.

Thank you to all the botanical adventurers, intelligent and intrepid women and men, whose voices down the years have told me how beautiful the world is.

Thanks to Fred Carotte and Laura Plume, my companions on this long voyage.

Many thanks to the magnificent Kew Gardens and its so welcoming crew.

Florence Thinard

ABOUT THE AUTHOR

Florence Thinard was born in 1962 in Royan on France's Atlantic coast. As a child, she exhibited a remarkable inability to do schoolwork anywhere near the ocean. The beaches, dunes, forests, ocean breezes, the ocean itself and all of nature called to her instead. She labored through school and eventually earned her baccalaureate. When she moved to Paris she finally found enthusiasm for studying and earned diplomas in history, political science and international relations.

After several years of observing the world through books, she decided to see it for herself and was a tour guide for groups in the United States, Turkey, Egypt, Thailand and other places. When she returned to France she became a print journalist, specializing in writing for youth and explaining the news.

For a decade, she has written documentary books as a way of satisfying her increasing curiosity about history and the way the world works. To balance this demanding work, she also writes young adult novels where reality and imagination intertwine. That way, she can set sail for distant horizons.

www.florencethinard.fr

FOREWORD

The plants are dried out, cut up, folded and refolded, twisted ... and yet so present.

The written notes are sketchy, almost illegible, crossed out, labeled, superimposed ... and yet so precise.

The herbarium specimens collected in this book can tell incredible stories all by themselves. Their botanical content tells of the explorers who labored to find and collect them. Their stains and folds speak of long, torturous journeys across continents and oceans. The many labels and notes they have accumulated reveal their extensive history of being studied by generations of scientists.

Are these herbarium specimens beautiful? Not always, because they are worn and heavily laden with their multiple stories. But that only makes them stronger, more evocative, almost more alive, because they reveal all the expeditions and adventures they endured before reaching us today.

Even though they are centuries old, they appear before our eyes thanks to the slightly-mad explorers who traveled all over the planet to collect them and bring them back. And thanks to the people at the Royal Botanic Gardens, Kew, in London and the Herbier de Montpellier, who continue to conserve, study and annotate them as our knowledge of the living world advances.

And thus, the herbarium continues to live.

Warmest thanks to everyone who carried them on part of their journey, and especially to those who allowed us access so that we could present them to you in this book: John Harris, Lydia White, Gina Fullerlove, David Goyder, Julia Buckley, Craig Brough, Véronique Bourgade and Peter Schäfer.

Frédéric Lisak,
Director, éditions Plume de carotte

THIS BOOK WAS CREATED IN PARTNERSHIP WITH THE ROYAL BOTANIC GARDENS, KEW
especially the Herbarium collections

The Royal Botanic Gardens, Kew is one of the world's greatest botanical gardens. Over one and a half million people a year visit its famous glasshouses, historic buildings and beautiful displays of plants from around the globe. As well as being an extraordinary garden, Kew is a world-leading scientific organization with extensive living collections, a herbarium, research laboratories, library, museum, art galleries and the Millennium Seed Bank, which holds over 2 billion seeds in safe storage.

Kew Gardens adorns the south bank of the River Thames in west London. The Gardens were designed by some of Britain's most famous landscape architects, and date back to 1759. Today the 320-acre site is of huge historical importance in terms of architecture, botany and landscape design. In 2003, UNESCO declared the Royal Botanic Gardens, Kew, a World Heritage Site. Kew's research is wide-ranging – encompassing plant diversity, conservation and sustainable development around the world.

The Herbarium was founded in 1853 and was originally formed from several private collections of dried plant specimens, including that of Kew's first director Sir William Hooker. The collection includes specimens from some of the world's most celebrated scientists and intrepid explorers including Charles Darwin, David Livingstone, Richard Spruce, Ernest Wilson and Joseph Hooker, to name just a few. There are currently 8 million dried plant and fungal specimens held here, 350,000 of which are "type" specimens — the specimen by which the name of a plant is defined. The collections continue to grow, with around 37,000 new specimens added each year through active research programs around the world and exchanges with partner institutions. The collections, which are arranged according the latest DNA research, are used every day for the study of systematics, micromorphology, biochemistry and molecular genetics as well as being the basis of many of Kew's conservation programs. New species of plants are named here almost every day. The correct identification of plant species is the foundation of all understanding of the plant world. Specimens are also constantly being digitized so that they can be shared with the world online.

The Library, Art and Archives collection is also vast and unsurpassed. In total it holds more than seven and a half a million items including books, art and illustrations, photographs, letters, manuscripts, periodicals and maps. The Library is one of the most important botanical reference libraries in the world. Kew cares for more than 200,000 prints and drawings, with many original works of art from the masters of botanical illustration including GD Ehret, PJ Redouté, the Bauer brothers and W.H. Fitch, as well celebrated contemporary artists such as Christabel King, Pandora Sellars and Rachel Pedder-Smith. The art collection is regularly shown in exhibitions in the Shirley Sherwood gallery of Botanical Art. Special collections include the works of Marianne North, which are housed in the gallery named after her in the Gardens and letters written by Charles Darwin just before and during his famous Beagle voyage. The Library collection dates from 1852, though items in the collection date from 1380 right up to the latest research papers, which are available to everyone.

Scientists and horticulturists alike use the specimens, artifacts and plant knowledge that have been collected at Kew to further the understanding of plants and fungi, to aid their conservation and protect the Earth's environment for the benefit of all.

You can access many of Kew's collections online — please go to www.kew.org, where you can also learn much more about Kew's work and the Gardens.

Christina Harrison,
Editor, *Kew Magazine*

You can see some of Kew's digitized collections and learn more about Kew Gardens and the work going on there at www.kew.org.

Lydia White, John Harris and David Goyder of the Royal Botanic Gardens, Kew

Without fear of contradiction, we can say that any herbarium exists because of the explorers. It grows and continues growing thanks to harvests, exchanges and observations by men and women from all times and places. They have explored backyard gardens, local streets, mountains on the horizon, the end of the world and the four corners of the Earth. The material they brought back remains an incomparable scientific tool, a database and a resource to be explored, concentrated in a very small space in a large world: a herbarium.

Some names are more familiar to us than others. Some are associated with exotic locales and infinitely rich continents. From the 17th century to the present, some of those who enriched the collection in the Montpellier Herbarium included Tournefort, Jacquemont, Delile and Monod. Others were explorers of the wide world, circumnavigating the globe in the 18th and 19th centuries: Commerson, Bougainville and Dumont d'Urville were among them. By bringing back specimens of flora and fauna from previously unknown regions, they furthered our discovery and understanding of the planet. New uses were discovered, whether ornamental garden plants or the economic development of new materials. By putting these samples into the collection, they built up a scientific heritage we still depend on today. But they were not the only explorers.

Many people have explored the environment closest to them. They are not as well known, perhaps, but they are just as fundamental to the understanding of life and they have contributed to discovering and naming the elements in their local heritage. In Montpellier, the great figures include Coste and his impressive herbarium and Candolle and Dunal whose research on Mediterranean fungi is still being used by mycologists. And there were many who patiently, passionately and anonymously added specimens to the herbarium. This work continues today, still adding specimens to tomorrow's herbarium. Collections are being enriched as new specimens are mounted and notes patiently inscribed on labels, recording a precious image of the world at a particular moment, in a particular place.

There is a final type of explorer, more discreet perhaps, but no less important for the understanding of our heritage: the passionate amateurs and researchers who work every day to use and enhance the value of this preciously conserved information. Using modern techniques, the essence of the information (in digital images and labels) is becoming accessible to the entire world, giving the collection a new dimension. By bringing the MPU Herbarium to life and enabling it to remain an exceptional heritage resource, and by making it known worldwide, they are explorers of yet another world; the world of herbaria.

Michel Robert,
President, Université de Montpellier 2

Véronique Bourgade and Peter Schäfer, Pôle Patrimoine scientifique (Scientific Heritage Center), MPU Herbarium

INTRODUCTION

VOYAGES TO PLANTS UNKNOWN

There were still many blank areas on Western maps in 1875: the high latitudes and the interiors of South America, Africa and Oceania.

WHY LEAVE?

Botany is not a sedentary and lazy science that one
can study in a dim and quiet office; it asks you to
journey through mountains and forests, climb rocky
cliffs and stand on the edge of precipices.

Fontenelle, Éloge de M. de Tournefort, 1709

Why would we leave our loved ones, our home and
our garden to take to the road and the sea, looking for unknown plants, when traveling, every
day is a trial. Century after century, the tales told
by botanical explorers are full of bitter comments
on accommodations. Instead of a bed, they had a
blanket spread on wet, vermin-filled straw, if not on
the bare ground or sand. Shivering, sleeping only
lightly, watching for rats, wild beasts and lurking
dangers and waking with a twisted back, going on
for another day of walking, through mud or snow
or blistering heat. "It is easy to imagine the feelings
of a European transported from a temperate climate
to the hottest place in the universe. My shoes grew
hard and shrivelled, then came apart, and finally
turned to dust ... and just the heat reflected by the
sand took the skin off my face," reported Michel
Adanson when he returned from Senegal in 1753.
They ate what they could glean from nature or
from the locals. Despite some pleasant surprises,
empty bellies were more frequent than feasts. Some
accepted this spartan life as a holy duty. In 1894, the
director of the Muséum in Paris wrote about Father

Armand David, "He had two blankets for sleeping
on a simple board and one bottle of brandy for
medicinal purposes; that was all he took for a year's
journey. The food he found in a Mongol tent or a
Chinese hut was always enough for him."
Among the greatest nuisances, after hunger, thirst
and illness, were the insects. Fleas, lice, moths and
worms have plagued unwashed humanity since
ancient times. Tropical insects added their exotic
torment. Deep in Central America, Humboldt and
Bonpland buried themselves in sand in an attempt
to escape the horrific mosquitoes, only to find themselves prey to carnivorous ants. George Forrest, who
explored the jungles of Burma in the early 20th century, described their attacks with dark humor:
"Creatures with inconveniently long legs plunge
suddenly into one's soup, great caterpillars in
splendid but poisonous uniforms of long and gaily
colored hairs arrive in one's blankets with the busi-
nesslike air of the guest who means to stay. Lady-
birds and other specimens of coleoptera drop off the
jungle down one's neck, whilst other undesirables
inset themselves under one's nether garments. The
light in the tent attracts a perfect army of creatures
which creep, buzz, crawl or sting."

Those who left were gone for a long time. When
Marco Polo's father left Italy in 1253 on his voyage
to China, his wife was pregnant. When he returned,

Sometimes the indigenous peoples were not welcoming. In 1787, 12 members of La Pérouse's expedition, including Lamanon, the botanist, were massacred by Samoans.

he met his son, who was 15 years old. Most travelers simply walked. The luckier ones had a horse to ride or a pack animal. Later, they crowded into rough wagons or stagecoaches, but the roads were barely passable. In 1815 it still took 4 days to travel from Paris to Nantes and 8 to go from Paris to Toulouse. Once they arrived at their destination, travelers were very isolated. In 1735 it took 2 long years for La Condamine's letter to reach the Académie des sciences in Paris and for him to get an answer back. Some explorers' letters expressed the sadness of leaving a wife and children at home.

Life at sea was even worse than life on the road. At the fringes of the known world, sailors faced innumerable dangers. What maps that existed were sketchy and inaccurate. An unexpected reef, a strong wind or a fire could lead to death. Whereas adventurers on the land faced highwaymen and bandits, those who traveled by canoe, caravel or full-rigged ship feared attack by pirates, enemy ships or cannibals in dugout canoes. The ship's crew had other worries than the comfort of scientists, who were relegated to leaky alcoves, along with their books, crates of equipment, moldy plants and animal remains awaiting taxidermy. Everyone was thirsty and everyone was hungry. On the larger vessels the ordinary diet was bread or biscuits, salt meat, salt fish, fresh water, beer and wine. Even if there was enough food, after weeks at sea it inevitably became rancid, moldy or rotten. The bread was full of weevils and the water was green with algae. Such a poor diet and the close quarters combined to encourage the spread of fatal diseases: typhus, dysentery, typhoid fever, and that terror of long-distance voyagers, scurvy, which left Captain Cook's crew "crawling about the decks ." In port, rats could carry plague and mosquitoes malaria. They never knew what welcome awaited them at landfall. On some Polynesian islands, women offered themselves to the sailors whereas on others, the natives fought them with fierce resistance.

From Magellan to La Pérouse, from Rumphius to Cunningham, more than one botanical explorer is buried in a faraway land, a victim of fever, shipwreck or spear wounds. Very few of them died in their beds. Even fewer of them were recognized for their heroism by lasting glory or fabulous wealth. So why did they go?

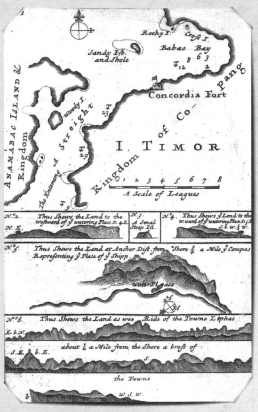

The first map of Timor drawn by William Dampier, privateer and scientist.

A New Zealand warrior in full dress. The Europeans were armed with diplomacy, sabers and rifles.

THE GRASS IS GREENER IN DISTANT PASTURES

It is impossible to conceive the original floral wealth of this country.

Ernest Wilson, *China, Mother of Gardens*

If we want to find new flowers and new trees, we have to leave home. There are more than 248,000 plant species in the world. Europe only has about 12,000 of them, whereas South America has some 165,000 species, Oceania 45,000, China 32,000 and India 21,000. Why is the vegetation so limited in Europe and so abundant elsewhere? The late Mark Flanagan, a botanist with the Royal Botanic Gardens at Kew and at Wakehurst Palace, unraveled this mystery: "The catastrophe for the ancient temperate forests of the northern hemisphere were the repeated glaciations in the Pleistocene. While the Earth was cooling rapidly, four glacial eras came in succession. The last one only retreated northward some 18,000 years ago, a mere blink in geological time ...

Humboldt, a geographer, and Bonpland, a botanist, at a rough campsite in the heart of the Amazon.

When the glaciers advanced, conditions in the higher latitudes worsened and the plants that grew there retreated toward the south ... In North America the forests' retreat was orderly because the mountain ranges oriented north-south were not an obstacle. At the height of the ice age, the temperate flora simply paused in Central America: it returned northward when the land thawed. In Asia, indications are that the glaciers were not as extensive and the temperate forests were not much affected. In Europe, on the other hand, not only were the glaciations severe and long-lasting, but the way south was less certain. Physical barriers made migration more difficult: the mountains formed an east-west wall from the Pyrenees in the west, to the Alps, to the Tatras in the east. If the plants got through these obstacles, they were up against the Mediterranean, with the mountains behind them. Only the hardiest and most adaptable species survived ... If your imagination is good enough, you may hear their dying screams echoing down the ages. The glaciers advanced and receded again and again. After their final retreat, the plants were able to re-colonize the north. In Europe the colonizing flora had been considerably reduced, and some key elements of the pre-glacial flora were missing. There were no more magnolias, no more pecans or sweet gums, katsuras, tupelos or catalpas — the list is as long and depressing as the roll call of soldiers who fell in a long-forgotten battle."[1]

Britain suffered even more, since the water levels had risen as the Earth warmed, and the British Isles had become isolated and unreachable by migrating plants. Its original trees number only 40 species. However, its cool, humid climate provided excellent conditions for the flora of temperate Asia. This advantage encouraged a great many plant hunters from the British Isles to bring the world's riches home to their splendid gardens.

1 *Plants from the Edge of the World,* Mark Flanagan and Tony Kirkham, Timber Press, 2005.

The Jamaican jungle, as painted by Marianne North. Tropical forests are an inexhaustible source of botanical discoveries.

WHAT WOULD EUROPE'S GARDENS BE LIKE WITHOUT THEM?

Without the botanical explorers, the variety of plants in Europe would be much reduced, and would lack color and perfume. Imagine the spring without camellias, tulips, hyacinths, mimosa, forsythia or lilacs. Imagine summers without trumpet vines, hydrangeas, oleanders or bougainvillea. No begonias or rhododendrons. Forget the fragrance of roses, gardenias, peonies or jasmine. Erase the color of dahlias, poppies or morning glories. Imagine the fall without the bright reds of Virginia creeper or maples, a park without birch or cedar, the streets without their plane trees or schoolyards without their horse chestnuts.

Spectacular tropical orchids would not have been discovered without plant hunters.

EAT, HEAL AND GROW RICH

The quest for spices began it. From the days when the Romans, in their journeys and their wars, first acquired a taste for the hot or aromatic, the pungent or intoxicating dietetic adjuvants of the East, the Western World found it impossible to get on without a supply of Indian spices in cellar and storeroom.

Stefan Zweig, *Conqueror of the Seas – The Story of Magellan*

Wild plants have been an essential source of nourishment since the dawn of humanity. Nomadic hunter-gatherers watched for edible roots, game and ripe fruits. Through unimaginable suffering, they discovered the healing, poisonous or hallucinogenic properties of many plants and domesticated them as well as they could during their travels.

About 8,000 BCE people began to harvest grains and legumes, plant them and keep the seeds for the next season. Thus, agriculture began in the Neolithic era on all continents and with it the desire to produce more and better harvests. This goal led to seed selection, improved varieties, studying and testing, and the exchange of seeds.

The first written trace of this great change comes from ancient Egypt. Hieroglyphics tell us that the kingdom of the two crowns was supplied with incense, gum, myrrh, red pigment and "all good smelling herbs" by the Nubian caravans that traveled east of the Nile. Fifteen centuries ago, the Pharaoh Hatshepsut was the first to initiate a search for plants

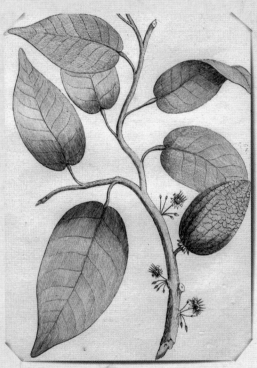

Cocoa pods on a cacao tree. This plant native to the Americas has become indispensable.

The secrets of tea cultivation were guarded by the Chinese for 5,000 years until Robert Fortune uncovered them in 1851.

for religious and strategic purposes. Myrrh was needed for embalming the dead and incense was indispensable to the priests because it wafted their prayers toward the gods. Sending her ships to find the living plants that produced myrrh and incense, Hatshepsut was working for Egypt's independence. Plants became an economic resource, as sources of food, medicine and wealth, and they retain that status today.

Hatshepsut, Marco Polo, Christopher Columbus, Magellan — they all shared an obsession with these prized plants: to discover where they came from, control the trade routes and secure the supply. Beginning in the 13th century, European adventurers, merchants and warriors in search of spices set out for Asia and the Americas. Because they were rare, pepper, cinnamon, nutmeg and cloves were worth their weight in gold and provoked international conflict, also leading to the birth of capitalism and the first multinationals: the Dutch, French and British East and West India Companies. But it was not only the greed of merchants that transformed plants into gold. Fortunes were lost overnight over tulip bulbs in the tulip mania of 16th-century Holland and over new orchid varieties at upper-class English auctions during the 19th century's orchid madness. Botanical exploration often followed in the footsteps of saber-wielding colonial armies and holy-water-sprinkling missionaries. The kingdoms and then the states of Europe competed in pushing back the frontiers of unexplored but promising territories. The stakes were huge. More than 80 percent of medicines came from plant sources at that time and burgeoning industries were seeking lucrative discoveries such as plants producing coffee, cocoa, quinine and rubber. In the 20th century, the struggle to possess plant resources shifted to industrial cultivation, seeking the best varieties of grains, tubers, fruit trees, grapevines, cotton and the like. Even the 21st century is not exempt, with competition over GMOs, exotic woods or new molecules hidden in old-growth forests.

THE THIRST FOR KNOWLEDGE

I'm crazy about botany: it gets worse every day. I no longer have only hay in my head, I'm going to become a plant myself one of these mornings.

Jean-Jacques Rousseau

Over the centuries many men, and some intrepid women, set out on adventures for the love of plants, to satisfy pure curiosity or to understand the world and solve its mysteries. In the great intellectual surge of ancient Greece, botany held pride of place. Aristotle and his disciple Theophrastus studied Egyptian and Babylonian knowledge and added their own contributions in scientific descriptions and rigorous analyses. But their ideas were lost in the shadows of the Dark Age after the fall of Rome when feudal barbarism reigned. All through the Middle Ages, ignorance and superstition ruled. Knowledge was hidden away in monastic scriptoria and withered. Luckily, at the same time, in Baghdad, Cordova and Cairo, Arab scientists were translating the Greek texts. They kept the knowledge alive. They taught it and added to it with their work in medicine and surgery, and in experiments with all kinds of plants.

During the Renaissance, the Greek and Latin texts were rediscovered through the Arabic translations.

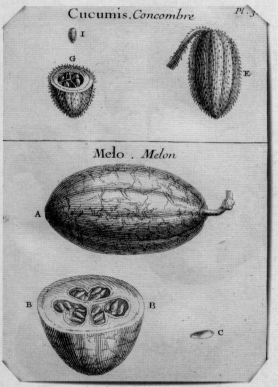

Since ancient days, drawing has been a precious method of recording plant characteristics.

Rosa centifolia bipinata painted by Pierre-Joseph Redouté

HOW TO NAME A PLANT

Linnaeus's nomenclature of 1753 replaced a system that relied heavily on each botanist's and gardener's memory. Before Linnaeus, plant names attempted to be descriptive, leading to names such as *Rosa sylvestris alba cum rubore, foglio glabro* for a briar rose or eglantine. After Linnaeus, every plant received a name composed of two words. The first is the genus and takes a capital letter, "*Rosa*"; the second indicates the species and is written with a small letter, for example, *Rosa gallica*, the French rose. This system was so effective that in English we know many flowers and plants by their Latin names alone. This binomial nomenclature has evolved to take account of hybrids (*Rosa x francofurtana*) and varieties (*Rosa gallica var. centifolia*). Linnaeus took many of his names from mythology (*Centaurea, Artemisia*, etc.), while other names honor the explorers (*Bougainvillea, Camellia, Banksia*...). Many species names evoke geography (*japonica, alpina*, etc.), physical characteristics (*minusculus, reptans, ferox*, etc.) or color (*azureus, rubens, versicolor*, etc.) or name the discoverer (*darwinii, plumierii*, etc.).

Once again, they flowed into Western universities and society. Botany rose from the ashes and became separate from medicine. Then the great work of naming and classifying plants began. People tried to impose order on the chaos of nature. Concerning the plants Tournefort gathered in the Middle East, Fénélon wrote, "Not everyone will understand that the pleasure of seeing them all, whole and well preserved, set out in proper order on big sheets of white paper, is sufficient regard for all they cost him." This vast organizational undertaking culminated in Linnaeus's stroke of genius, to identify each plant using only two words, the family name (genus) and individual name (species).

After the discovery of America, interest in the natural world intensified. Curiosity cabinets and rooms became all the rage in Europe. Nobles, scientists and wealthy amateurs bought many things from travelers: shells, rocks, stuffed animals, rare and strange objects, and exotic plants, of course. They displayed them in glass-fronted cabinets or filled rooms with them. The general public went to botanical gardens to look at rare plants. The garden in Pisa opened in 1543, Montpellier in 1593 and the royal garden in Paris in 1635.

During the 18th and 19th centuries, traveling naturalists developed a more systematic approach. They began taking inventory of known and unknown worlds, and their work was written down and illustrated by the encyclopedia creators of the Enlightenment. For the first time in history, exploration was not all about conquering and gaining wealth, but also discovery for the purpose of knowledge. Buffon, the intendant of the royal garden, named himself the "correspondent for the king's curiosity cabinet" and thus received first choice of the most beautiful and remarkable specimens discovered. He sent explorers out upon the seas, including Louis-Antoine de Bougainville, Pierre Sonnerat, Joseph Dombey and Jean-François de La Pérouse. These brilliant scholars and gentlemen of the 18th century were able to embrace the whole scope of scientific knowledge in their time. Geography, astronomy, physics, botany, zoology, mineralogy, geology, anthropology — they were acquainted with and curious about every part of science. Their work was done in extremely arduous conditions and has lasted an exceptionally long time. Their maps, astronomical observations and descriptions of the people, plants and animals have been used for centuries and remain invaluable references.

A 17th century curio cabinet, filled with an assortment of animal skeletons and objects from the New World.

The largest seed in the world is a botanical curiosity: the double coconut.

Better yet, these wise scholars were able to cross boundaries and fraternize with each other for the good of science and humanity. Even when their countries were at war, French, English, Prussian and Russian botanists were exchanging herbarium specimens, rare live specimens and letters describing their discoveries. One example of such scientific pacifism is the journey of the crates containing La Billardière's Australian harvests. In 1793, after a long and winding journey, the crates belonging to this French botanist and revolutionary landed at Kew Gardens and ended up in the hands of Joseph Banks, an English royalist. Banks, as a gentleman, refused to take "one single botanic idea away from a man who discovered them at the risk of his life" and returned the crates to their owner, intact.

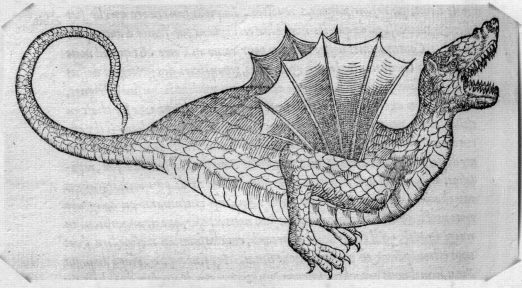

A dragon engraved by Pierre Belon (16th century). The early explorers reported on all of nature's marvels, both the ones they saw and the ones they heard about.

All captains and masters of merchant vessels from Nantes sailing to foreign countries and France's American colonies shall be expected to bring seeds and plants back when they return.

Royal decree, 1726

Discovering new plants is one kind of exploit but bringing them back in good condition is another. With rare exceptions, the early explorers settled for fruits, seeds, pips, pits, bulbs, rhizomes, cuttings and even grafts for fruit trees. Roots and leaves were dried, and sometimes reduced to powder.

Herbarium specimens composed of dried plants, first known as *hortus siccus* or "dry gardens," began to appear in the 16th century. They made it possible to observe the physical features of a plant — its roots, stems and flowers — and compare it to others and classify it. Of course, the herbarium specimens had to get home safely, and many years of hard work and suffering were reduced to nothing by shipwrecks, piracy, theft and fire.

It wasn't until the great voyages of discovery in the 17th and 18th centuries that it became

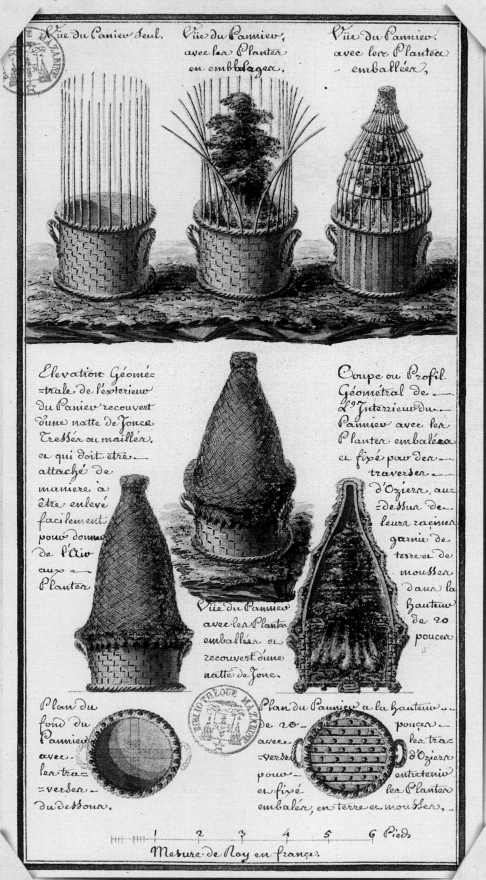

Despite royal orders and elaborate packing baskets, the 17th century saw immense numbers of plants lost.

more common to transport living plants. In 1753 Duhamel du Monceau, inspector general of the French navy, wrote the first practical guidebook, *Advice for the Transportation of Trees and Live Plants, Seeds and Other Natural History Curiosities by Sea.* In it he recommended numbering the plants, cataloguing them under their French names, their foreign or "savage" names, and their "true names and characteristics" and recording "whether the fruits were edible or useful for medicine or the arts."

One of the botanists' challenges was to convince the captain to leave the plants on deck, where they hindered operations, instead of putting them in the hold. Deep in the hold, they were in the dark with brine around their roots. Since salt is the enemy of plants, they had to be rinsed after every big spray, but fresh water was rare at sea. Botanical history recounts the sacrifice of the Chevalier de Clieu who gave his own water ration to a coffee plant while crossing the Atlantic.

The plants traveled in pots or in grilled crates that protected them from rats. The packing process improved. Shrubs were placed in tall wooden crates with air holes and metal corners that resisted the ship's rolling and lurching. Within the crates, the roots were wrapped in good, thick layers of earth and moss. Smaller plants, or those traveling by land, were sheltered in reed baskets woven around their foliage. These baskets were covered with rush mats that had to removed to give the plants some air and put back if there were storms or a drop in temperature.

Despite all these precautions, the losses were enormous. A plant from the "torrid zones" has only one chance in 100 of resisting the cold near Europe. In France, the survivors were taken into one of the five "port gardens" created in Nantes, Rochefort, Lorient, Brest and Toulon at Buffon's request. Damaged plants were revived before be-

ing shipped by barge to Paris. The biggest revolution came in 1830 in a London doctor's conservatory. Wanting to preserve his ferns against the dense London fog, Nathaniel Ward shut them into sealed glass boxes. Surprise! Not only were they safe from urban pollution, but they flourished without being watered. The good doctor sent two empty cases to Australia, asking that they be returned, full of plants, as quickly as possible. Barely 2 years later, he received species that had never before survived the long journey. The experiment was so conclusive that Ward's terrariums were soon an absolute necessity for plant hunters. Joseph Hooker used them first in Oceania and Robert Fortune sent his stolen Chinese tea plants to the Himalayas in Wardian cases.

Seeds today are still washed, dried and annotated. Botanists still press plants in newspaper to create herbarium specimens (Théodore Monot's preferred papers were Le Monde and Libération). Today, in addition, specimen cells are now stored in gels that keep them fresh enough for DNA testing.

Ward's watertight and airtight terrariums, known as Wardian cases, revolutionized the transportation of live plants.

Father Pierre d'Incarville sent thousands of seeds from China to the Jardin du roi, wrapped in thick paper.

Basalt prisms in a waterfall, sketched by Humboldt. Naturalists were often scholars of all sciences, including botany, zoology, geography and geology.

DISCOVERY, WONDER AND ANGER

The greatest pleasure for a naturalist, although more practical persons may not think so, is discovering new species, putting new islands on the map of nature and sometimes even populating apparently empty continents.

Richard Spruce, 1851

It can be disturbing to imagine white men crossing an ocean, getting off their boats , picking up a plant and declaring that they have discovered it, even though it has been known to the local population for hundreds or thousands of years. Generations of men and women had eaten it or used it as medicine, weapon or dye. But such ancestral use is not dependent on a scientific description. By "discovering" we mean "recording the existence of a plant by sending its scientific description, a drawing or a specimen to a place where they will be conserved and become a universal reference." Some naturalists make the point that their plant discoveries are "new to science."

Beyond the lure of novelty, botanists still revel in the beauty of the world. It is difficult to imagine the cultural and sensual shock when they first encountered the temperatures and landscapes on the far side of the world. In 1786 La Pérouse described the island of Maui as delightful. "We could see waterfalls tumbling down the mountainside into the sea, after providing water for the natives' homes ... You would have to be sailors like us, reduced to one bottle of water a day in searing heat, to truly appreciate the sensations we felt. There were trees on the mountaintops, banana trees around the houses, and everything stirred our senses with its charm."

Their reports are dazzled odes to nature, to the abundance of the forests and the exuberance of the flowers. In the tropics the fragrances breathe and colors vibrate. Who better than a painter to evoke this sense of wonder? Marianne North, freshly arrived in Jamaica in 1871, wrote, "There was a small valley at the back of the house which was a marvel of loveliness, bananas, daturas, and great *Caladium esculentum* bordering the stream, with the *Ipomœa bona nox*, passion flower, and *Tacsonia thunbergii* over all the trees, giant fern-fronds as high as myself, and quantities of smaller ferns with young pink and copper-colored leaves, as well as the gold and silver varieties. I painted all day..." and later, "I passed one evening through a cleft in the rocks so narrow I could touch them with either hand; they were covered with a scarlet lichen, pretty green and purple orchids growing among the moss. The allspice-trees were showing their white flower-buds and the leaves were very sweet when crushed."

Botanists were lovers of nature and often showed themselves to be pacifists and humanists as well, despite the spirit of the times that insisted on the superiority of the white "race." Living for long months with the indigenous people, many botanists grew to like them most sincerely. Joseph de Jussieu and the Peruvian Indians, Michel Adanson and the Africans in Senegal, and Carl Linnaeus who owed his life to his Lapp friends, were among many who spoke out, publicly and in writing, against the poor treatment of colonized peoples. "In our opulent cities where our luxury requires products from the four corners of the world, we enjoy all the gifts of nature without considering what procuring them is costing humanity," wrote Pierre Poivre in 1768. And his thoughts are just as valid today!

Sadly, the naturalists' worries about the plundering of the environment are equally true in our modern day. Among many others, Father Armand David, a missionary in China in the 1870s, said, "One is sad to see the speed with which the primitive forests are being destroyed; in all of China, they are in tatters and will never be replaced. When the great trees disappear, a multitude of shrubs and other plants that can only flourish in the shade go with them, along with all the large and small animals who need forests to live and reproduce ... And unfortunately, what the

China's exceptional flora has been pushed back for millennia by the quest for agricultural land.

Chinese are doing in China, others are doing elsewhere. It is truly sad that humans have not developed enough in time to save so many other beings that the Creator put on this earth to live alongside humans, not only to decorate this corner of the world, but also to play a useful and necessary role in the world's economy."[1] A hundred years later, in 1984, Margaret Mee made the same observation about the Amazon basin, "The landscape (such as it remains) is a dismal expanse of capoeira (scrubby jungle land) and where the virgin forest used to stand, there is only a blackened lake full of huge skeletons." In 2012 the virgin forests, the most wonderful reservoirs of biodiversity, are in ruins. Botanist Francis Hallé would add, "It is an immense reservoir of biochemical molecules, a planetary treasure chest that opens up vast opportunities for pharmaceutical research. One day we will need these molecules and we'll say, 'That was stupid. We had them in our hands and didn't use them?'"[2] At the age of 95, after a life spent crossing and studying deserts, he proposed a new way to explore and said, "It may be useful, sometimes, to sit under a tree and look around ourselves. We must teach children to look at the living details of an insect or a flower. That will bring us back to the same level as other living beings."[3]

1 *Le Nuage et la Vitrine, Une vie de Monsieur David* by Emmanuel Boutan, Éditions R. Chabaud, 1991.
2 *Interview with Télérama.* October 2008.
3 *Terre et Ciel*, Actes Sud, 1997.

Pierre Sonnerat, a curious and attentive observer, makes a study of a parrot in New Guinea.

PORTRAITS OF THE
BOTANICAL EXPLORERS

MYRRH AND INCENSE FROM THE LAND OF PUNT

To the greater glory of Pharaoh Queen Hatshepsut

The first female pharaoh and the fifth ruler in the 18th dynasty, Hatshepsut ruled from 1479 to 1458 BCE. Daughter of Thutmose I, she married her half-brother Thutmose II, who died young. His son, Thutmose III, came to the throne as a child. Hatshepsut named herself pharaoh and was represented as a man with a beard. Her 20-year reign was stable and prosperous for Upper and Lower Egypt. "The Two Kingdoms bent to her will and served her," wrote the mayor of Thebes. Hatshepsut had her mortuary temple complex built at Deir el-Bahari in the Valley of the Kings. Her name was later erased from her tomb, probably by order of Thutmose III.

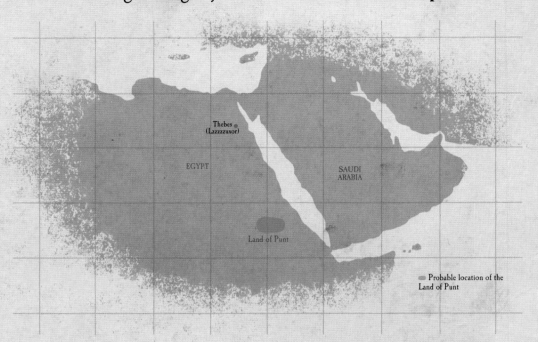

Probable location of the Land of Punt

A mun himself, the ruler of the gods, "commanded that the marvelous products from the Land of Punt be brought to him, because he loves Queen Hatshepsut even more than all the other kings who have ruled this country." Thus began one of the most fabled voyages in Antiquity, recounted in the exceptional bas-reliefs of the temple at Deil el-Bahari in the Valley of the Kings at Thebes.

That divine instruction came at an opportune time for the Pharaoh Queen Hatshepsut, because border problems were interrupting the arrival of caravans from the Orient. The Egyptians needed supplies of myrrh and turpentine, two of the resins used in embalming, and in frankincense, the indispensable incense whose smoke carried the words of the gods. Hatshepsut ordered the construction of five strong ships with white linen sails, propelled by 30 rowers. Counting sailors and soldiers, the expedition consisted of 1,000 men under the command of the chancellor, Nehesy. They set off in the year VII or VIII (around 1465 BCE) for the Land of Punt, "land of the Gods." The hieroglyphic record explains that these adventurers crossed a great body of water, the "Great Green," but the exact location of Punt remains a mystery: was it Somalia? Eritrea? Sudan? Yemen? Did they go up the Nile or down the Red Sea? The bas-reliefs show the people of Punt resembling Egyptians except for the beards, and huts on stilts surrounded by date and coconut palms, baboons, leopards and a giraffe.

Welcomed warmly, the Egyptians gave gifts, met the king and queen of Punt, then loaded their "ships very heavily with the marvels from the land of Punt; all goodly fragrant woods of God's Land, heaps of myrrh-resin, with fresh myrrh trees, with ebony and pure ivory, with green gold of Amun, ... with incense, eye cosmetic, with apes and dogs ..." Thirty-one incense trees of the genus *Boswellia* were transported in baskets filled with earth. The precision of the engraving makes it possible to recognize their massive trunks, their narrow, oval leaves, and the small irregular drops of resin exuded by the bark.

The return to Thebes was a triumph. The incense trees were replanted in front of the temple at Deir el-Bahiri and other treasures offered to Amun and his priests, while the people sung the glories of divine Hatshepsut.

AN IMMORTAL GARDEN

When Hatshepsut died, Thutmose III took power and undertook a dozen military expeditions that made Egypt the center of a vast empire. On his return from a campaign in the Near East in year 25, he established a "botanical garden" of 275 exotic and extraordinary plants. Irises (*Iris albicans* and *oncocyclus*), low cornflowers (*Centaurea depressa*), Abyssinian euphorbia, lemon balm and myrtle all bloomed around the great temple of Amun at Karnak. The pharaoh explained, "... a fertile land produces its fruit for me. My Majesty has done this so they are offerings to my father Amun ... now and forever."

Bas-relief from the temple of Deir el-Bahari depicting the loading of the incense trees and gifts from the people of Punt on board the Egyptian vessels

"Her majesty herself held out her arms and spread the incense all over her body. Its fragrance was like the perfume of the god, its odor combined with that of Punt and her skin, now gilded like finest gold, shone as bright as the stars in the palace of festivals."

from the bas-relief of Hatshepsut's tomb from the French translation by Sylvie Griffon (Cestas, France) © 2010 Sothis-Egypte

FACING PAGE

HERBARIUM PLATE

Boswellia carterii

This specimen of frankincense was collected in 1875.

Möhr méddhu Hildebrandt
= Mohr madow Birdwood
Specimen agrees well with figure of
B. Carterei in Linn. Trans XXVII t.29.

Boswellia Carteri Birdwood

1381. No.

vernac. Möhr méddu (méddu = schwarz).

Statio Somali-Land: Meid.
Ahl- u. Serrudgeb. 1000-1800 m
In Ritzen d. Kalkfelsen.
arbor 4 m. alt.
Mutterpfl. d. echten Weihrauchs.

Leg. J. M. Hildebrandt.
com. Rensch. April 1875.

FOLLOWING ALEXANDER'S FOOTSTEPS TO AN EXOTIC HARVEST

Theophrastus, botanist of the Greek empire

Theophrastus was a nickname meaning "divine speaking ability"; his real name was Tyrtamus. He was born in Eresos on the Greek island of Lesbos around 372 BCE. His father was a fuller and in that workshop he learned to use plant-based dyes. A good student, he was sent to the Academy in Athens where he met Aristotle, who taught about understanding the world through observation. From 347 to 343 BCE, they worked together to collect information on fauna and flora, and Theophrastus used this material as the basis of his *Enquiry into Plants*, which was a collection of notes for the new course he taught at the Athens Lyceum. He was Aristotle's successor as director of that prestigious institution and remained in that position until his death in 287 BCE.

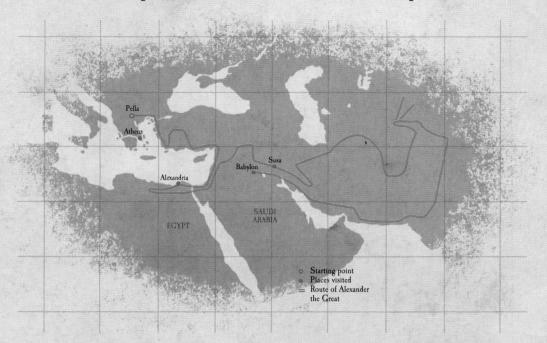

Pella
Athens
Alexandria
Babylon
Susa
EGYPT
SAUDI ARABIA

○ Starting point
● Places visited
═ Route of Alexander the Great

In the year 334 BCE Alexander III, king of Macedonia, led 30,000 foot soldiers and 5,000 mounted men to conquer the King of Kings, Darius of Persia. Only 22 years old, he was beginning one of the greatest epics in the history of humanity. In 329 BCE he conquered the provinces of Persia and Bactria (Afghanistan) and in 327 BCE he invaded India's Punjab. What might have been a simple military conquest became an outstanding exploration of natural science because of certain people among Alexander's companions. These included Aristobulus the architect, Callisthenes the historian, Androsthenes of Thaso the biologist, and Nearchus and Onesicritus, naval officers. Theophrastus was known as a man of great intelligence, but since he was too old to go on the expedition, he asked them to describe every detail of these foreign lands, and their flora and fauna. His "collectors" sent him letters and sometimes seeds and fruit.

Back at the Lyceum in Athens, Theophrastus compiled, experimented and theorized, and classified their discoveries using a scientific method that would not be equaled for 1,500 years. His harvest of exotic plants reflected the outsized scope of the adventure. From Egypt came knowledge of the sycomore, acacia, date palm and sensitive plant. From Libya came the jujube tree and the cypress, from Persia the pistachio tree, from Babylon came seeds of the citron in 331 BCE. The first of the citrus fruits introduced into the Mediterranean basin, this "golden apple tree" was the biggest success for a millennium. India was the land of all novelties. Theophrastus was enthralled by the mangrove trees, the banyan tree whose open-air roots made "a sort of tent where one can while away the hours," the banana tree with its ostrich-feather

HISTORIÆ
PLANTARUM
UNIVERSALIS
Tomus III.
Auctoribus
JOHANNE BAUHINO,
Archiatro.
JOH. HENRICO CHERLERO,
Doctore:
Basilienfibus
Quam recenfuit et auxit
DOMINICUS CHABRÆUS,
D. Genevenfis,
Juris verò publici fecit,
FR. LUD. A GRAFFENRIED,
Dñs in Gerkenfee
EBRODUNI,
MDCLI.

Cover of the Historia plantarum, *1651 edition*

leaves. He argued in favor of the economic potential of cotton, rice and spice plants, such as cinnamon, cassia, balsam, spikenard, etc. Onesicritus, who had met Brahmins, also described the mango tree as "a tall tree with wonderful, large fruits, on which fed Indian wise men who wear no clothes."

In 325 BCE Alexander retreated and lost half his men in the terrible desert of Balochistan, while Nearchus was taking the imperial flotilla home across the Persian Gulf, with great difficulty. But on the shores of Arabia, the sailors made one last, precious discovery, "an ample provision of myrrh and incense with which they filled their vessels before setting sail."

GARDENS OF BABYLON

The famous Hanging Gardens of Babylon created by Nebuchadnezzar in Iraq, a hot, sunburnt land, were one of the seven wonders of the ancient world. Alexander was overwhelmed and ordered that plants be transported there from Greece, especially those that provided dense shade.

TOXIC DISCOVERIES

Alexander's soldiers risked their lives testing unknown plants. Aristobulus warned against the tamarind tree, "which has pods like beans, ten fingers long and full of honey: it is not easy to save someone who has eaten of it."

"They saw vines appear in India on the mountain called Meru from which, as we know, legend says Dionysius came."

Theophrastus of Eresos

FACING PAGE

HERBARIUM PLATE

Gossypium herbaceum and *Gossypium indicum*

These specimens of cotton were harvested in 1834–1835.

Gossypium herbaceum L.

(Cephalonien) cultivirt an flachen
Stellen bei Lixuri. Erndte im
august & Septr. gesammelt
den 29. octbr. 1834.
Schimper & Wiest
miserunt Aprili 1835.

Culta prope Lixuri Cephaloniae.
U. i. d. 29. Oct. 1834. Schimper et Wiest.
Hooker & Wiest miserunt Jul. 1835.

G. herbaceum L.

G. herbaceum, Linn.
Wild & Cult. Cottons, p. 158
J. Watt

ARABIAN BOTANY IN THE TIME OF THE CRUSADES

Ibn al-Baitar, the Andalusian in the Orient

Abu Muhammad Abdallah Ibn Ahmad Ibn al-Baitar Dhiya al-Din al-Malaqi was born in Malaga around 1190, when that southern port city in Spain belonged to the immense Islamic caliphate of al-Andalus. This seat of culture attracted a great many scholars including Ibn al-Baitar's father, who was a renowned veterinarian. The young Ibn al-Baitar studied with famous physicians and botanists in Seville. They helped him collect plants in Spain and taught him to observe, experiment, analyze and classify. Ibn al-Baitar later traveled across North Africa and took up a position with the sultan in Egypt, whom he accompanied to Damascus, where he died in 1248.

Malaga · Constantine · Tunis · Tripoli · Damascus · Cairo

○ Starting point
● Places visited
— Route

In 1219, Ibn al-Baitar, age 21, left his native Andalusia to collect medicinal plants on the North African coast. His travels took him to the lands we know as Morocco, Algeria, Tunisia and Egypt, and as he traveled he accumulated many plants and observations. For instance, he was the first to describe the argan tree and methods of extracting and using its oil.

When he arrived in Egypt, he found a country at war. Sultan al-Kamil was trying valiantly to resist the Christian invaders who were besieging Damietta. The young botanist became friends with the sultan and was appointed chief herbalist in Egypt. In 1221, Sultan al-Kamil succeeded in negotiating the crusaders' withdrawal by ceding Jerusalem to them. Taking power over Syria, his court settled in Damascus in 1227. Ibn al-Baitar undertook a new hunt for medicinal plants in Palestine, Asia Minor and the Arabian peninsula. A man of amazing intellect and a researcher with wide-ranging interests, he was the first to discover a therapy for cancerous tumors (a herbal mixture called Hindiba) and the first to take an interest in weeds that were a problem for farming and classify them according to the crops harvested. He also studied the chemistry of rosewater and orange blossom water, observed both marine and terrestrial fauna, and extracted essences from animals and metals.

Ibn al-Baitar spent the years 1240–48 in Damascus, writing his masterwork, *Kit̄ab al-j̄ami'li-mufrad̄ at al-adwiya wa al-aghdhiya*, known in English as the *Compendium on Simple Medicaments and Foods* or the *Book of Medicinal and Nutritional Terms* or the *Complete Book of Simple Medicaments and Nutritious Items*. It contained all the pharmacological knowledge of his time, referring to the work of 150 Muslim doctors and 20 Greek scientists. He described

Preparing a remedy. Plate from De materia medica *by Dioscorides*

300 new plants and reported on his observations in Andalusia, the Maghreb and the Orient. He methodically classified 1,400 plants in alphabetical order with their names in Berber, Arabic, Persian, Syrian, Latin or Castillian whenever possible. He listed watermelon, cumin (good for digestion) and *Nigella sativa* (black seed), which he said was "a cure for every ailment but death." The mandrake, whose root roughly resembles the human form, provided a rare moment when he abandoned his scientific rigor; he said it was a remedy for epilepsy and for "all the maladies caused by genies, demons and Satan." Although the great book was not translated into Latin until 1758, it was used as a reference by Arab scholars all through the Middle Ages.

THE ARABS KEPT KNOWLEDGE ALIVE

While Europe suffered barbarian invasions and cultural pursuits were confined to monasteries and scriptoria, science and philosophy were flourishing in the Muslim world. Translations into Syrian and Arabic saved many Greek manuscripts from oblivion, including the thoughts and work of Aristotle, Hippocrates and Galen. Thus, thanks to Ibn al-Baitar's commentary on Dioscorides' *De materia medica*, the work of that first-century Greek doctor, a pharmacological compendium including plant illustrations, can still be studied.

"These carrot seeds are known in Syria as qomaylae and hasisat al-baragit around Jerusalem, because people mix the seeds with perfumed oil, spread them on their beds, and the fragrance of the seeds puts fleas to sleep and prevents them from biting."

Abu Mohamed Abdallah ibn Ahmed al-Baitar Dhia Ad-Din al-Malaqi

FACING PAGE

HERBARIUM PLATE

Argania sideroxylum Röm & Schultz

This specimen from an argan tree was harvested in 1823 in a garden in Perpignan, France.

THE SILK ROAD AND ITS SPICES
Marco Polo in the court of Kublai Khan

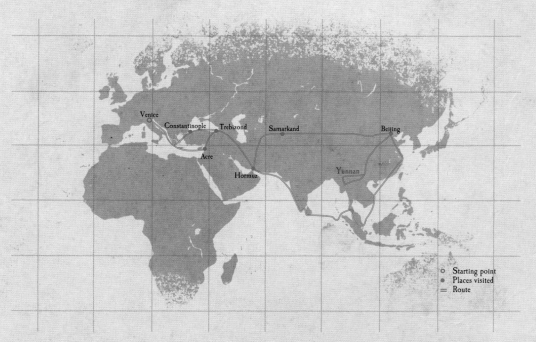

Marco Polo was born in 1254 into a family of Venetian merchants. His father, Nicolo, and his uncle Maffeo had traveled to the court of the great Mongol ruler, Kublai Khan. When they returned Marco's mother had died and the young Marco was 15. He was 17 when the three of them left for China in 1271. Marco became very useful to the great Khan and served as his special envoy. Laden with riches, the Polos returned to Venice in 1295 after a 24-year absence. During the constant battles between Genoa and Venice, Marco was taken prisoner and dictated his travel stories to his cellmate, a writer named Rustichello, creating the *Book of the Marvels of the World*, now known as *The Travels of Marco Polo*. After his release, Marco married and had a family, dying in Venice at the age of 70 in 1324.

○ Starting point
● Places visited
= Route

In the 13th century, Venice was the greatest merchant city in Europe. Its traders supplied luxurious goods from the Indies. Marco Polo's father and uncle, Nicolo and Maffeo, had already traveled to China in 1253. Not without difficulty, they took the silk route that went north of the Caspian Sea. The emperor of the Tatars, Kublai Khan, welcomed them warmly and even asked them to ask the pope to send missionaries to him. But when they returned to Rome, a new pope was about to be elected. The Polo brothers waited 3 years, and decided to leave in 1271, this time taking Marco, who was 17.

The first stop was Acre, where the family had a business office. From there, "to avoid the dangers of the roads and evil events of wars," they went on horses through "little Armenia" (Anatolia), the land of Turks ("ignorant and rude people"), greater Armenia and Persia (whose "inhabitants are nasty, quarrelsome, crooks, thieves and assassins"). After walking for days "little water is to be met with, and that little is impregnated with salt, green as grass" and bitter, they arrived at the Persian Gulf. Then their route took them north through Afghanistan, Kashmir, western China and the terrible Gobi Desert, haunted by "evil spirits." That part of the trip was done in winter, and the Polos were often stopped by snow and floods. Happily, "Kublai Khan, having heard of their return, sent more than 40,000 of his people to meet them."

While traveling, Marco Polo, true to his merchant upbringing, estimated the possibilities. Here there were gold and turquoises, there, "dates, pistachios, paradise apples (bananas)," and elsewhere, "pearls, golden cloth, silks, velvets, ivory and other precious goods." The young traveler paid particular attention to spices. During the 17 years he spent in the service of Kublai Khan, and then on his return through Indonesia, Ceylon and India, he never stopped looking for the sources of spices. In China, there was "a root called rhubarb" and "great quantities of ginger," in Tibet "much cinnamon and other fragrant spices," in Cochinchina, "aloe wood and ebony forests," in Java "an abundance of pepper and nutmeg." *The Book of the Marvels of the World* is full of these descriptions that fascinated Marco Polo's contemporaries and, 200 years later, sent Christopher Columbus out onto the Atlantic Ocean.

A copy of the Book of the Marvels of the World *or the* Devisement du monde, *with 265 paintings by the Master of the Mazarine, was prepared for the Duke of Burgundy around 1410–12. In this picture, pepper is being harvested in India.*

THE BRAZIL OF THE INDIES

In Sumatra, Marco Polo found brazilwood (lignum brasilium). Imported from the Indies, its red bark was used as a dye. Brazil is named for this wood, because the brazilwood of the Americas, Pernambuco (*Caesalpina echinata*) grew there in abundance.

THE EMPEROR'S PAPER MONEY

Marco Polo was sometimes accused of telling lies. Venetians could not believe in the money used by Emperor Kublai Khan, "which is neither gold nor silver," but the fine inner bark of the mulberry tree, cut into pieces and embossed with the Emperor's seal.

"There are cloves in great abundance, gathered from trees with small branches and white flowers that produce a fruit, in which the cloves appear like the seed."

Marco Polo

FACING PAGE

HERBARIUM PLATE

Cinnamomum verum

This specimen of cinnamon was first collected in 1831 and was refurbished several times prior to 1974.

Palmicollah April 1835

Cinnamomum verum Presl

DET. Kostermans 1972

FLORA OF MADRAS
Det. J. S. Gamble
1914

2511

Peninsula Indiae Orientalis
No.

Cinnamomum zeylanicum Nees
Herb. Wight.
Distributed at the Royal Gardens, Kew. 1866-7.

SAILING WEST TO THE SPICES OF CIPANGO

Christopher Columbus, conqueror of the Atlantic

Christopher Columbus was born in 1451 in Genoa, and like many Genovese, found his life's work on the sea. He worked on sailing ships on the Mediterranean, and then between Portugal, England, Iceland and Madeira. He settled on Madeira and traded a great deal with the Canary Islands and Azores. In 1484 he was proposing to "reach the east by going west." The king of Portugal refused to finance his voyages, but Isabella of Castile financed four of them. The first two (1492–93 and 1493–96) took Columbus to the Caribbean islands. On the third, (1498–1500) he reached Venezuela but thought it was an island. His fourth voyage (1502–1504) ended with a shipwreck and a year of being stranded in Jamaica. Wealthy but weak and out of favor, Columbus died in Valladolid at 55, in 1506.

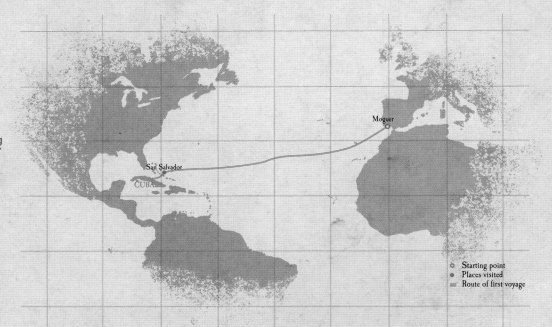

- ○ Starting point
- ● Places visited
- ═ Route of first voyage

Is the earth round? Yes! was the answer given by many European scholars and sailors as early as 1480. Christopher Columbus was one of them. An experienced merchant and an outstanding navigator, he had studied Marco Polo's *Book of the Marvels of the World* and made many notes in his copy, as well as the Catalan maps of the coasts of India, China and Japan (known then as Cipango). Based on faulty calculations Columbus convinced himself that the spices and riches of Asia were within reach of the brave sailor who took an Atlantic route. The king of Portugal rejected his idea in 1484, so Columbus offered it to Queen Isabella of Castile. At first she was not interested but she was eventually convinced by the idea of profit. She paid for two caravels, a carrack (a four-masted ship) and a crew of 90 ruffians and adventurers. On August 3, 1492, the flotilla left the port of Palos de Moguer, headed west. The crossing was difficult and uncertain, "to the mysterious rim of the Occidental world."[1] Rations ran short and the dead calm seas frightened the crew.

But at last, on the morning of October 12, Columbus set foot on the island he named San Salvador, then continued to the coasts of Cuba and Hispaniola.

That first expedition was one of conquest. Columbus wrote brief, technical descriptions of the land, but his amazement showed through. "On the island named Juana I also saw seven or eight kinds of palm trees whose majesty and beauty, and that of all the other trees, plants and fruit, easily surpass our own," he noted in his log as he sailed along the coast of Cuba. On that first

trip he brought back some plants, including peppers, which he so named because he was sure he had landed on "the Indian islands beyond the Ganges."

The goals of his second voyage, from 1493 to 1496, were trade and colonization. Two doctors traveled with him to answer the two essential questions, "Where are the spices?" and "What will the colonists eat?" They found corn, "a kind of millet that comes to a point and is nearly the thickness of one's upper arm," and yucca that the "Indians" used to make bread. They recognized aloes, lemons, cherry plums and turpentine trees, but could not hide their disappointment: the cinnamon "was not as good as that found in Spain," the tiny dates "were only good for feeding pigs." Columbus brought back a little gold, some slaves, pineapples and tobacco. The results were poor compared to expectations. Despite two other expeditions, the Admiral of the Ocean Sea died before he reached the American continent, and never knew its immensity or its wealth.

OFF BY 1,000

Columbus estimated that China would be about where Florida is. He was relying on a correct estimate of the Earth's circumference, made in ancient times, but his error lay in his interpretation of the figures used by the Arab scholars who had translated Greek and Latin texts using a "mile" of 6,473 feet (1,973 m) and not the Roman "mile" of 4,862 feet (1,482 m).

TRANSATLANTIC SUGAR CANE

As a trader, Christopher Columbus regularly purchased cane sugar from the Portuguese at Madeira. In 1493 he introduced roots of *Saccharum officinarum* to Hispaniola, from where it spread to Sao Tomé, Cuba, Jamaica, Brazil and Mexico.

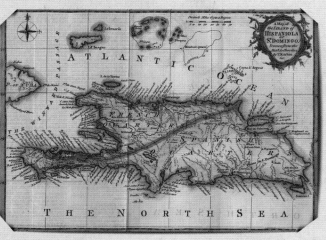

A map from 1758 of the island of Hispaniola

1 José Maria de Hérédia, *"Les Conquérants" (Conquistadors)*

"I promise our invincible rulers, who have given me some help, that I will give them as much gold as they require, as much spice as they desire along with cotton and gum ... and aloe wood, and as many slaves as they request."

Christopher Columbus

Nicotiana tabacum
macrophylla
under cheese cloth.

ECONOMIC PLANTS OF THE WORLD
Distributed by C.F.Baker

No. 2. Nicotiana tabacum L. var. macrophylla Schk.
      Collected March 1, 1907, at Santiago de las Vegas,
Cuba, by C.F.Baker. Illustrating the effect of
growth under cheese-cloth. From these plants the
finest wrapper tobacco is taken. They become 5-8
feet or even more in height, and the leaves, though
not greater in number, become very much larger,
thinner, and finer, and are at much greater dis-
tances on the stem.

51

Gonzalo Fernández de Oviedo y Valdés was born in Madrid in 1478. His education by the Duke of Aragon was filled with Renaissance humanism. At 13 he was made a page to Prince Juan who was the same age, and they became friends. Thus, Oviedo was a witness to great events at an early age, and he began to write about them, for instance the siege of Granada in 1492 and Columbus's return from his first voyage. When the prince died at 19, Oviedo sought consolation in travel. He worked briefly as a secretary for the Inquisition, but at 36 left for America as a gold inspector and historian. He did not return permanently to Europe until 1556 and died at Valladolid, Spain, aged 75.

CONQUERING A NEW WORLD OF BOTANY

Oviedo and the Americas

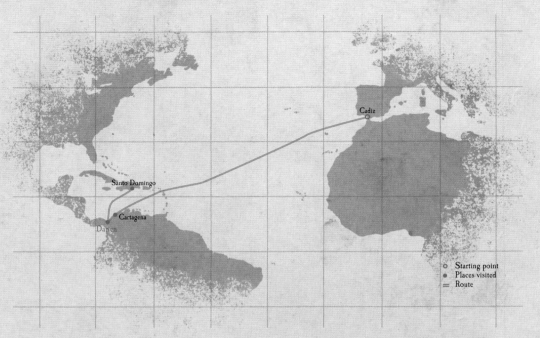

Cadiz

Santo Domingo

Cartagena

Darien

○ Starting point
● Places visited
= Route

"What mortal intelligence could comprehend ... such an infinite multitude of trees, those the Indians cultivate and those that Nature produces without the aid of human hands? So many plants that are useful to Man? And so many others that are unknown to us, and how many roses and flowers and fragrances?" Fascinated but not at all daunted by the immensity of the New World, Gonzalo Fernández de Oviedo y Valdés, known to us as Oviedo, threw himself into an immense project, his *General and Natural History of the Indies, Islands and Mainland of the Ocean Sea*. Although he was the official historian of the Catholic monarch, Isabella, his research and writing conditions were not comfortable.

In 1514 the Spanish were still struggling to settle on the Caribbean islands and the coast of the American continent. Oviedo arrived as the inspector of the Castile gold mines, in a region that covered modern Nicaragua and Costa Rica. He sailed with the armada of the conquistador Aria de Avila, who commanded 20 ships and 2,000 battle-hardened men eager for wealth. Gold excited their greed and that of privateers who attacked the mines. In addition, Avila turned out to be violent; he tyrannized the Spanish as much as the Indians. In 1520, Oviedo had to escape se-

cretly in order to tell the king. He returned in 1524, with a mission to establish order in the province of Cartagena where the conquistadors were destroying each other. After another voyage to Europe, he returned as the governor of the city of Santo Domingo, a frequent target for Caribbean pirates.

Despite that, Oviedo continued his gigantic botanical harvest and started rigorously classifying all the plants. His task was all the more difficult because, faced with such absolute novelty, he lacked the words to describe the Americas' bounty. He frequently had to use native names or make comparisons to European species. "There is a plant called ajes, which to my eye looks a little like a turnip from Spain." He was the first to attempt description of dozens of plants including potatoes, banana trees, pineapple, lignum vitae, barberry fig trees (which he called a monster tree) and *Cereus repandus* or Peruvian apple cactus. He noted the Indians' useful inventions such as the hammock and canoe, but found fault with others: "One of the bad habits of the Indians is to inhale the smoke they call tobacco. I do not understand what pleasure they get from it." That was only one of the insoluble mysteries of this unknown continent, so different from Europe.

Drawing of a pineapple by Fernández de Oviedo y Valdés, from his History, Historia general y natural de las Indias, 1851

REVOLUTION

Following in Oviedo's footsteps, Francisco Hernandez, the physician to Philippe II, led the first European scientific expedition to the Americas. In 7 years in Mexico, from 1570 to 1577, he described 2,500 species; a botanical revolution for the European continent that only knew 600.

TRANSATLANTIC IMPORTS AND EXPORTS

Oviedo was also interested in the trees brought by the colonists: orange, apple, pomegranate, date and olive trees, and grapevines; they all flourished better in the soil of the Americas than in Spain.

"The Indians eat human flesh and are abominable and cruel sodomites; they shoot arrows poisoned with herbs so that the wounded only survive by a miracle or die while eating their own flesh or the soil."

Fernandez de Oviedo

CLOVES THAT COST EVERYTHING
Magellan in the Moluccas (now the Moluku Islands)

Ferdinand Magellan was born around 1480 into an old noble family of northern Portugal. At 24, he joined the fleet that sailed to the East Indies on behalf of King Manuel I. Between 1505 and 1513 he was in every battle and fought to take Malacca alongside his friend Francisco Serrao. When Serrao went to settle in the Moluccas around 1511, Magellan became involved in a military campaign in Morocco. Since Magellan's reputation had suffered because of some lapses in discipline, Manuel I rejected his idea of a new route to the Moluccas. Magellan sought help from Charles V and began his expedition in 1519. He did not reach the Moluccas, nor did he sail around the world, because he died in the Philippines in April 1521.

○ Starting point
● Places visited
= Route

On August 10, 1519, the flagship *Trinidad*, with the *San Antonio*, *Santiago*, *Concepción* and *Victoria*, set sail from the port of Seville with 256 crewmen. Their leader was a young Portuguese officer, Ferdinand Magellan. Only 39, he had succeeded in convincing the young king of Spain, Charles V, to finance a risky venture: sailing around South America to reach the Moluccas. These islands, the source of cloves, belonged to the Portuguese, but Magellan planned to correct the maps of the Indies and give possession to Spain.

The expedition began well but soon fell victim to bad luck and mortal dangers, as vividly recorded by the knight, Antonio Pigafetta. The fleet spent its first winter in Patagonia, but in that icy desert, three captains mutinied. The rebellion was quelled but the *Santiago* was shipwrecked. Then, exploration of the strait that now bears Magellan's name went on for a month, as it was a labyrinth, "surrounded by high, snow-covered mountains." At Easter 1520 the desperate crew of the *San Antonio* took over the ship and deserted.

When the surviving sailors finally reached the vast Pacific Ocean, they cried tears of joy, but then faced 3 months and 20 days without fresh food. "We ate biscuit, which was no longer biscuit, but powder of biscuits swarming with worms ... It stank strongly of the urine of rats," recounted Pigafetta. The water was putrid and stinking. Plagued by scurvy, they survived by eating leather pieces of the sails. Finally, they reached the Marianas Islands and ate their fill of coconut bread and palm wine.

When Magellan landed in the Philippines, he required the islanders to recognize the rule of the King of Spain and pay him tribute in the form of gold, pepper, cloves, cinnamon and nutmeg. One chief named Cilapulapu preferred to fight and killed Magellan with his lance, on the beach on Mactan Island, on April 27, 1521. Magellan never saw the Moluccas, but his surviving ships got there in November. Their persistence was rewarded with large cargos of cloves, obtained at ridiculously low prices. But the *Concepción* had to be scuttled because there were not enough crewmen and the hull of the *Trinidad* was coming apart. Although the *Victoria* was leaking in many places, she miraculously rounded the Cape of Good Hope. On September 8, 1522, the 18 weakened survivors reached the Andalusian coast. Selling the cloves barely covered the expedition's costs. Henceforth, humans knew the limits of their world.

CHEAP CLOVES

In the Moluccas, Magellan's companions traded for their cargo of cloves with red cloth, hatchets, glass cups and ribbons, and even obtained "100 pounds for two bronze chains" worth 10 sous in France.

LIVING LEAVES

Pigafetta observed some very strange trees, "which produce leaves which are alive when they fall, and walk... I kept one of them for nine days in a box. When I opened the box, that leaf went round and round it." It was probably a leaf insect of the Phylliidae family.

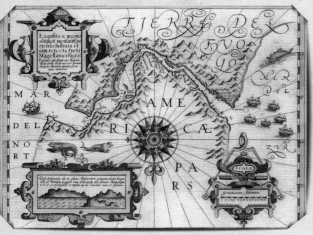

Map of the Strait of Magellan by Jodocus Hondius, 1606

"There is a passage leading from the Atlantic to the Indian Ocean. Give me a fleet and I will show it to you. I will go around the world, going from east to west."

Ferdinand Magellan

Bojer.

Eugenia cariophyllata.
in Moluccas.

IN THE NAME OF THE TULIP
Busbecq in the empire of Suleiman the Magnificent

Ogier Ghiselin de Busbecq, born in Flanders in 1522, was a faithful subject of Ferdinand I, Holy Roman Emperor and King of Bohemia and Hungary. He was an able diplomat and had served the emperor in England before being sent on a mission to the Ottoman Empire to negotiate peace with the irascible Suleiman the Magnificent. It took him 8 years of groveling, from 1554 to 1562, before he obtained the precious treaty. His adventures, as told in *Turkish Letters*, was one of the best sellers of the Renaissance, with 23 editions in Latin. Despite his impatience to return home to taste the pleasures of private life, Busbecq was posted to Holland and then France, where he died in 1591.

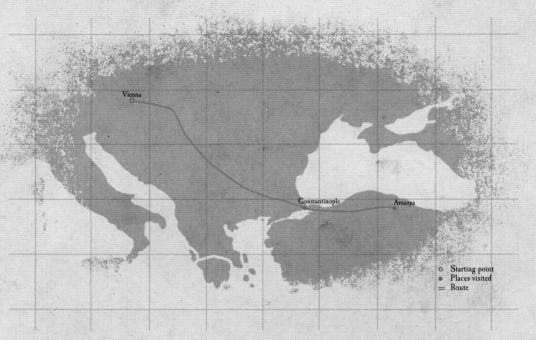

- o   Starting point
- •   Places visited
- =   Route

In the early 16th century, the Christian kingdoms were very worried about their southern frontiers. The formidable Ottoman army had besieged Vienna since 1529. The Europeans sent many emissaries to deal with Suleiman II; the Hapsburg emperor Ferdinand I sent Ogier Ghiselin de Busbecq. In 1544 Busbecq began his journey with an tumultuous coach ride, followed by stormy sailing on the Danube to Constantinople (now Istanbul), on the way to Amasya in eastern Turkey, where Suleiman was holding court. When Busbecq was finally granted an audience, he was sent away abruptly with the words "Giusel, giusel" meaning "Well, well," as soon as he had delivered his message. He was given a truce of 6 months to take Suleiman's reply, "wrapped up in cloth of gold and sealed," to Vienna. The ambassador hastened across the Balkans and returned with a mandate to negotiate a peace treaty with the Turks. He then spent 6 difficult years in Constantinople, more or less confined to his residence and subject to many frustrations.

The pacifist Busbecq admired the love shown by the Turks for flowers, which they cultivated with the greatest care, and bought bulbs from the Janissaries who guarded his door. Among these bulbs were tulips, which originate in central Asia on the slopes of the Himalayas. They had become familiar in Constantinople around 1000 BCE and bloomed in the gardens of Topkapi Palace, illuminated like jewels by lantern light. The Turks liked their tulips tall and thin, with sharp, pointed petals and lyrical names like Rose of Colored Glass, Light of the Mind, or Those that Burn the Heart. The gardeners created more varieties and tulips were everywhere at the height of the Ottoman Empire, embroidered on clothes, engraved in stone, and painted on tiles and pottery.

The bulbs Busbecq sent across the Bosporus arrived in Holland and ended up in the hands of Carlos Clusius, a famous Flemish botanist. He hurried to acclimatize them in his garden in Leiden. They were highly coveted and one night his garden was robbed and his bulbs stolen. That was the beginning of tulip mania that gripped Holland and spread throughout Europe. Tulips, especially those with multicolored petals, were worth at least as much as gold. In 1635 a single bulb of *Semper augustus* was sold for 13,000 florins, the price of a house with garden and stables on Amsterdam's most beautiful canal.

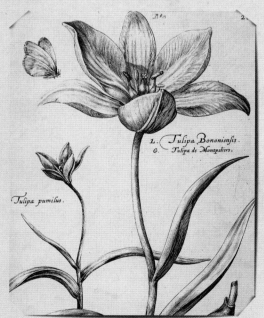

Engraving of Tulipa Bononiensis *and* Tulipa pumilus *(or humilus) from the book* Hortus Floridus, *first edition of 1614. The engravings were the work of Crispijn van de Passe the younger.*

THE TURBAN CONFUSION

The Turkish word for tulip is *lale*, but Busbecq speaks only of *tulipan*. The Turks often wore tulips attached to their turbans, so it is possible that Busbecq asked his interpreter what "that" was, but got an answer referring not to the flower but to the headwear, *tulband*.

THE ALCHEMICAL VIRUS

The capricious tulip bulb may suddenly produce feathery petals striped in red and white. This mystery was worth its weight in gold and was unsolved until the 1920s, when it was found that a virus was responsible for these sought-after anomalies.

"We saw meadows of the most beautiful variegated colors. The paths were bordered by narcissus, hyacinths and tulips, and a thousand Turks stood there, offering us bouquets all the way."

Ogier Ghiselin de Busbecq

FACING PAGE

HERBARIUM PLATE

Tulipa alpina,
Tulipa montana
and *Tulipa gesneriana*

These specimens of various species of tulip were harvested in 1834 and 1856. Our gardens also owe a debt of thanks to Busbecq for the common lilac and horse chestnut.

Tulipa alpina Gay mst. n° 152.
(19 Mart. 1856)
= *T. pulchella*, Fengl.

Balansa dabat
16 Jan. 1856.

Région alpine du Taurus oriental, au-dessus
de Bouïgarmaden.

Fl. L juillet. — fr. août.
1855.

Tulipa alpina — montana
Lindl. et cusicae Boiss. et Held.
quasi intermedia (J. Gay. 19 Mart.
1856)

Tulipa gesneriana d.

M. AK-Bag (Mont Blanc) Taurus oriental.
Juin 1854.
Montpeer Redit.
Febr. 1856.

C5
Bolkardagóli

400

THE ORIENT: LILIES, LILACS AND FORTRESSES

Pierre Belon, naturalist and spy?

Pierre Belon was born near Le Mans in 1517 and grew up in Brittany, already fascinated with flora, birds and fish. He was a gardener, an apprentice apothecary, and then a student of botany in Germany, thanks to René du Bellay, bishop of Le Mans and brother of a renowned poet. Belon traveled through Europe, studied medicine and then became apothecary to the bishop of Tourrion. A fervent Catholic, he wielded a vicious pen against the reformists and appeared to be devoted to the crown, serving François I, Henri II and Charles IX successively. After 3 years in the Orient, Belon wrote a noted history of his travels and several works on botany and zoology. A controversial, self-taught figure, Belon was mysteriously assassinated in the Bois de Boulogne in 1564.

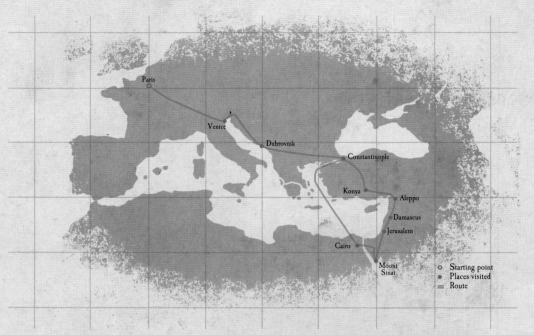

Starting point
Places visited
Route

Officially, Pierre Belon went to the Orient to observe the plants described by Theophrastus in their natural habitats. He traveled as the botanist in the ambassadorial mission sent to Suleiman the Magnificent by François I, who was seeking desperately to ally himself with the Turks against Charles V. The political situation was extremely confused in those years: alliances were ephemeral; Christianity was torn apart by the Reformation; and the Turks occupied the Hungarian plains.

Thus the French delegation led by the Baron d'Aramon left Venice in secrecy in February 1547. When they arrived in Croatia, Belon left the expedition to wander around the Greek islands and Crete for several months, barely escaping pirates and Turks, "who never kill the men they capture, but sell them instead." He crossed the Dardanelles, always taking copious notes on plants, animals and ruins, and also on well-guarded Turkish forts "equipped with good artillery." At Constantinople, Belon was the first to describe tulips, calling them "Turkish lilies." He took an interest in the market stalls, discovering one seed that made people sleep and another that made them happy. He was enchanted with the lilacs and sent cuttings to France in the diplomatic pouch.

He traveled all over Macedonia, then returned to Constantinople to meet the new ambassador, Baron de Fumel, who had been appointed by the new king, Henri II. Together they left for Egypt, reaching Alexandria and going up the Nile to Cairo. Belon described garden irrigation, the doum palms with their forked trunks, sycamores with their "exquisite foliage," black myrtle that "likes to live beside the sea," cassia trees and sycamore figs, and balm, that "famous, precious and rare plant." The ambassador

Illustration of a palm tree from the Herbarium Amboinense, 1750 by Georg Rumphius

and the botanist crossed the desert to reach Mount Sinai, walking at night to avoid the "violent heat."

Belon made many more observations and drawings of giraffes, chameleons, crocodiles, black ibis, Alexandrian mustard and Levantine acacia. On the way to Syria, they visited Palestine and the Holy Land, crossed Lebanon and climbed Mount Taurus. There, Belon left his traveling companions and spent the winter in Anatolia, a region unfamiliar to Europeans. Was he doing military reconnaissance on Suleiman's territory or was it a simple naturalist's expedition? If he was a spy, he was never exposed as one. His botanical legacy, however, is still present in the gardens of France: lilac, plane-tree, Judas tree (redbud), cedar, holm oak, olive trees and oleander.

MUMMIES

Visiting the Pyramid of Gizah, Pierre Belon was fascinated with the way bodies were preserved in Egypt. He learned that "mūmiya" did not mean the embalmed body, but the "chemical known to the Greeks as pissasphalto" or bitumen, which made conservation possible.

MAGNETS

Sailing to Egypt, Belon was thrilled to discover lodestone, once called Hercules's stone, which enabled a person to "set out in a small ship, in all kinds of wind and weather, and cross the sea."

"It was already very late when we found a stream running toward Aleppo. Having passed the stream, we left the soft earth behind and entered a stony land of mountains and rocks. We began to see trees bearing olives, pears, plums and almonds."

Pierre Belon

FACING PAGE

HERBARIUM PLATE

Platanus orientalis L.

This specimen of plane-tree was collected in Italy, on the eastern shore of the Adriatic, in 1869. This species was introduced into Europe by Pierre Belon and hybridized with the western plane-tree to produce the common plane-tree.

8

Platanus orientalis L.

Cannosa à 10 milles au N.
de Raguse. Individu ayant
9 mètres de circonférence à
1 mètre du sol.
Ch: Martins 3 mai 1866
Magazin pittoresque 1870. Dec. 1870

P. Or. var. cretica Dode
Determinavit   P. Rivas

Raguse, East coast of Adriatic Sicily
(another R. in Sicily).

IN SEARCH OF RARE PLANTS

Tradescant the Elder in Russia and Algeria

John Tradescant the Elder was born in eastern England around 1570. His talents as a gardener led to his employment by Robert Cecil, Earl of Salisbury and a minister to Elizabeth I. Later he worked for the Duke of Buckingham, whom he accompanied on a failed expedition to relieve La Rochelle in 1628. During his long career as a plant-hunter, Tradescant also gathered natural curiosities and displayed them in his London house, called The Ark, establishing the first museum open to the public. His son, John Tradescant the Younger, was also a gardener and collected plants on his three voyages to Virginia.

Arkhangelsk (Archangel)

London

Formentera

○ Starting point
● Places visited
═ Route of first voyage
═ Route of second voyage

Nearing the age of 50, John Tradescant the Elder had a solid reputation as a gardener to England's aristocracy, and was also a tireless traveler and an experienced merchant. For a decade he searched the nurseries of France and Holland looking for original and rare plants for his employer, the Earl of Salisbury. Tradescant designed and improved the earl's gardens, constantly buying new varieties of tulips, narcissus, lilies, iris, roses, anemones, asters, buttercups, cherry trees, pear trees, apple trees, quince trees and hawthorns.

When a British embassy was sent to the Czar of Russia, he took this opportunity to broaden his botanical horizons. The long journey went around Scandinavia to reach Archangelsk. Tradescant made few observations of the voyage in his journal, except for noting the ambassador's constant seasickness and the thick fog near the Arctic circle. Arriving in the delta of the Dvina in August, Tradescant wanted to be taken "from island to island to see what was growing on them." He discovered an unknown rose, which he named *Rosa moscovita* and at that time revealed that he had no sense of smell. Although his nose was not good, he truly had a green thumb. Despite the cold and the lack of fresh water, he returned to England with beautiful carnations, white hellebore and mercury, in addition to his rose.

He had barely returned when he set off again, in 1620, on an expedition to "crush the Barbary pirates" of the Mediterranean. For many years Algerian corsairs had been harassing British shipping, taking crews as slaves and capturing the cargo. The expedition was a military fiasco, but that mattered little to Tradescant who went ashore in two botanist's paradises, Malaga and Formentera. He may also have gone ashore in Algeria because he described "the fields of Barbary covered with gladioli." The gardener brought back from these sunny lands roots of starry clover, sarsaparilla, terebinth, white and pink rockroses, and a fine wild pomegranate with double flowers, all wrapped in damp moss.

In 1624 Tradescant rose in stature once more, becoming the right-hand man of the Duke of Buckingham, the very wealthy favorite of James I. His highest honor came in 1630 when he was appointed head gardener to Charles I. His later travels were limited to escorting the queen consort, Henrietta Maria of France, known as the "Rose and Lily Queen" for the alliance of the Tudor and French crowns.

THE CABINET OF CURIOSITIES

Whether it took up a single cabinet or several rooms, a curiosity cabinet was intended to sum up the world and penetrate the secrets of Nature. In the 16th century, such cabinets contained a multitude of rare specimens of animal, vegetable, mineral or manufactured items. The emphasis was on the fantastic, abnormal and bizarre. Tradescant's Ark held not only exotic plants, but also two whale ribs, a stuffed chameleon and a stuffed pelican, a number of petrified objects, a banana, a monkey's head, shells, a mermaid's hand, precious medals, a fragment of the True Cross and a hatband made of snake bones.

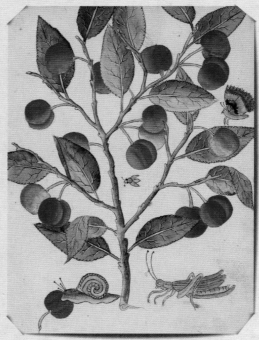

Watercolor of fruit and additional fauna from The Tradescants' Orchard, *1620*

"I also saw shrubs of all sorts ... the wild roses, single, in a great abundance, spread over four or five acres ... They look a great deal like our cinnamon rose; ... those who have the sence of smelling say they be marvelus sweete.»

John Tradescant the Elder

Punica granatum
M. Orte pone M. S.te Lugi
1822 Mai

HERBARIUM
1854
BENTHAMIANUM

HERBARIUM
1854
BENTHAMIANUM

"*Punica granatum*"
Hort Audibet 1828

HERBARIUM
1854
BENTHAMIANUM

Magna Vern
J. Colchester 1832

SERVING THE FIRST MULTINATIONAL SPICE COMPANY

Georg Rumphius, Soldier and Botanist in the Moluccas (Moluku Islands)

Georg Eberhardt Rumpf, known as Rumphius, was born and raised in Germany, where he received an excellent education and also learned Dutch from his mother. In 1646, at the age of 19, he went to fight in Brazil, a territory in dispute between the Dutch and Portuguese. But his ship was captured and Rumphius had to spend several years in Portugal. In 1649 he returned to Germany to assist in his father's business, and eventually joined the Dutch East India Company. Sent to the Moluccas in 1654, he began as an officer and became a merchant, also exploring the local flora and fauna. Although he went blind and experienced many tragedies, he continued his study of nature until he died in 1702.

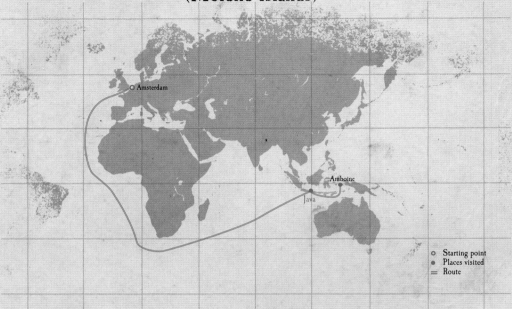

- ○ Starting point
- ● Places visited
- ═ Route

A new form of capitalist business was born in Amsterdam in 1602: the corporation. Known to the English-speaking world as the Dutch East India Company, the United East India Company (in Dutch, *Vereenigde Oost-Indische Compagnie*, or the VOC) set out on the route to the Indies to wrest the flourishing spice trade from the Portuguese. Financed by its shareholders and given a royal charter to operate as it pleased, the VOC soon had a substantial fleet of ships that could both trade and fight. It owned a network of private fortified ports reaching all the way to Japan. This first company to operate worldwide opened a branch in Java and gained the upper hand over commerce between Europe and Asia, especially in the trade between Ceylon, India, China, Japan and Indonesia. The VOC bought and sold millions of florins worth of pepper, silk, pearls, cotton and sandalwood, among other commodities. It created a monopoly on trade with the Japanese market and another on the finest spices such as nutmeg, cloves and cinnamon.

When Georg Rumphius, a ship's officer, landed in the Moluccas in 1654, his primary goal was to ensure the VOC's control over cloves. The company had limited the cultivation of this spice to the island of Ambon by removing the plant from all the other islands in the archipelago. Rumphius was fascinated by the flora and fauna and quickly became a trader for the VOC. His goal was to find remedies for the tropical diseases that struck colonists by studying the local plants and knowledge of them. At 42, he lost his sight, likely due to glaucoma, but despite this "sad darkness" he continued his research. He had a prodigious memory and relied on others to write down his words and create illustrations. In 1686 he faced another disaster: his wife and one of his daughters were killed in an earthquake. The next year, a fire ravaged the European quarter and destroyed his manuscript and all the drawings he had made before going blind.

Nonetheless, the first six volumes of the *Herbarium Amboinense* were completed in 1692 and sent by ship to Holland to be printed. But the ship was sunk by the French. Stubbornly, Rumphius went back to his task, assisted by his surviving daughters, and finally, after 30 years of labor and disasters, 12 volumes, with 2,000 plant descriptions and 811 illustrations were delivered to the VOC's headquarters. But the company did not publish the work because of the huge costs of printing and because it considered the information too sensitive. And so Rumphius died in 1702, without seeing his life's work published.

Illustration of a palm tree from the Herbarium Amboinense, *1750 by Georg Rumphius*

THE PLINY OF THE INDIES

In addition to the 12-volume *Herbarium*, Rumphius also wrote *The Ambonese Curiosity Cabinet*, covering shells, corals and birds. The extent of his works earned him the nickname "the Pliny of the Indies," and they are still regarded as a treasure by botanists, anthropologists, ethnologists and chemists.

UNRECOGNIZED

Although Rumphius was the first to describe many plants, Linnaeus attributed them to other botanists. Still, his name remains on several species of palms — *Caryota rumphiana* or fishtail palm, *Cycas rumphii* or queen sago palm, *Calamus rumphii* and others — and orchids (*Coelogyne rumphii*).

"I beg of you to excuse me for not writing very much about the clove tree, or how it is seeded and cultivated, because that is expressly forbidden."

Georg Rumphius

Coll. Soejatmi

No. 345   7. IX. 1975

Fam. Palmae

Gen. Licuala

Spec.

Det.

Vern.

Island South Sulawesi

Loc. near Pakalu, Maros

Habitat

                                   Ca 20 m. alt.

Notes   Not common in dist.
         below the rocks.
         3 m tall. Fr...
         to red.

Licuala m...

...ore,
...ey, New
...bane, Can-
...st. Bogor.
...cation of this specimen

DRAWING THE FLORA OF THE CARIBBEAN

Charles Plumier, monk and artist in the Antilles

Charles Plumier was born to a modest family in Marseilles in 1646. Keen on mathematics, he entered the austere order of the Minims and his studies took him to Rome where he developed an interest in botany. He studied under Pitton de Tournefort and went herb-hunting in Provence. Then he studied painting, engraving and lens making. Asked to participate in making an inventory of the flora of the Antilles, Plumier sailed across the Atlantic three times. He became the royal botanist and over 15 years wrote 22 volumes of botanical descriptions with wonderful drawings. He died near Cadiz in 1704 as he was about to set out on his fourth voyage, which would investigate the Chincona tree.

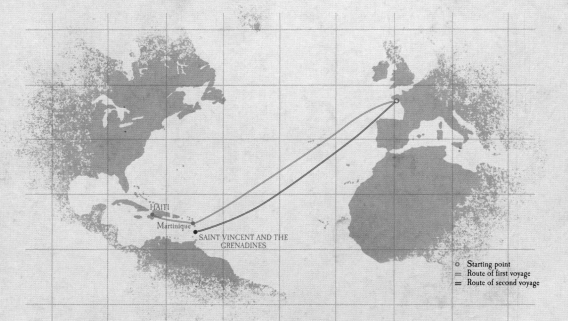

HAITI

Martinique

SAINT VINCENT AND THE GRENADINES

o   Starting point
=   Route of first voyage
=   Route of second voyage

In 1689, Louis XIV ordered an inventory of the natural resources of his Antilles: Guadeloupe, Martinique and Santo Domingo (now Haiti). Michel Bégon, intendant of the royal navy, recommended sending the famous chemist, Dr. Surian, who suggested inviting Charles Plumier, in whom he saw "a wonderful talent for botany and a fine hand for drawing plants." The monk's quiet life took an unexpected turn. Although Plumier wrote little about their adventures, his traveling companion, a Dominican friar named Labat, wrote with a sharp and elegant style. Labat wrote about Surian and Plumier, "They left in 1689 and came back 18 months later, laden with seeds, leaves, roots, salts, oils and other things, and a great many complaints about each other."

Plumier returned to France with only his notes and drawings because his herbarium and specimens had traveled on another ship and were lost at sea. Still, his work was of such high quality that Louis XIV appointed him the royal botanist and also financed the engraving and printing of his first book, *Description des plantes de l'Amérique* with many splendid line drawings, done from life and often life-size. There were ferns (Plumier's favorites), plants in a new genus he named *Saururus* "because they resembled a lizard's tail," arums, peppers, clematis, passion flowers and more. He made notes of their curative properties but his frankness betrayed him. Labat chortled, "I have rarely known anyone easier to trick than this good monk." He recounted gleefully that the natives had convinced Plumier, who was terrified of snakes that might lurk on the savannahs where he was collecting, that a certain vine would chase the reptiles away. The skeptical Labat shut a snake and a piece of the vine in a glass jar. The furious snake bit into the plant without any ill effects. The poor botanist, who

had already drawn the vine and written up his medicinal notes, lamented loudly.

On his next voyages, in 1693 and 1696–97, he went to other Caribbean islands, in particular the Grenadines. Although suffering greatly from malaria, he described vanilla, cacao, cotton and begonias, named after his dear friend Bégon. In fact, it was Plumier's idea to give new botanical genera the names of explorers, doctors or naturalists, and also named fuchsia for Leonhart Fuchs, lobelia for Mahias de l'Obel, magnolia for Pierre Magnol, etc.

Engraving of various plants. From Description des plantes de l'Amérique, avec leurs figures, *Charles Plumier, imprimerie royale, 1693.*

MODESTY REWARDED

The humble father Plumier never gave his own name to a plant, but as an homage, Pitton de Tournefort and Linnaeus named the genus *Plumeria*, the fragrant frangipani, for him.

BESTIARY OF THE ANTILLES

Charles Plumier did not limit his studies to plants, and left more than 1,200 zoological drawings, some of them in color. With extraordinary precision he drew birds, snakes, lizards, turtles, fish, shells, bats and even a crocodile that he had captured so that he could study its jaw up close.

"Of all the plants I have discovered on the islands of America, there are none that give me greater pleasure than the ferns."

Charles Plumier

FACING PAGE

HERBARIUM PLATE

Fuchsia triphylla L.

This specimen of three-leaved fuchsia was harvested in Santo Domingo in 1910 by H. von Türckheim. Plumier's herbarium was lost at sea.

b

Fuchsia triphylla L.

Determined by P.E. Berry 1980
Missouri Botanical Garden

2 6 OCT 1910

FLORA VON SANTO DOMINGO.

Nr. 2936. *Fuchsia triphylla* L.

Prope Constanza 1250. m. a. margine
sylvarum

II 1910   Coll. H. von Türckheim.

THE FIRST FLOWERS FROM AUSTRALIA

William Dampier, scholar and pirate

Although his parents were modest English farmers, William Dampier received a good education. At 16, seeking travel and knowledge, he signed on with a merchant ship. At 20, he began keeping a journal as he worked briefly as a soldier, then a plantation manager in Jamaica and a logger in Mexico. By 1679 he had become a buccaneer and sailed around the world, gathering riches and surviving shipwrecks. He recounted his adventures in *A New Voyage Round the World*, which was a great success. Dampier went around the world three more times and, despite the rough life of a pirate, published other notable books. He retired in London where he died in 1715.

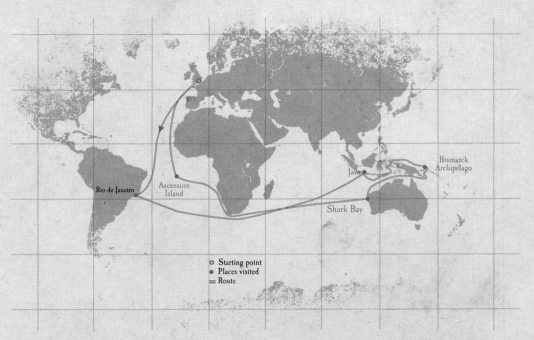

Bismarck Archipelago
Java
Shark Bay
Rio de Janeiro
Ascension Island

○ Starting point
● Places visited
═ Route

A sailor who loved his travels, a writer who loved to write about them, a pirate to fill his coffers, and a hydrographer, climatologist and botanist because of his love of nature, William Dampier had the soul of an adventurer and a fine scientific mind.

At the turn of 18th century, he was able to go around the world four times, but only once for a legal purpose. The British Admiralty was so impressed by his first circumnavigation, which took 12 years and made him the first Englishman in Australia, that it gave him a ship, the *Roebuck*, and a mission. His goal was to explore New Holland (Australia), New Guinea and the Malaysian archipelago. The voyage started badly; Dampier was an experienced pirate but a poor captain. Suspecting his first lieutenant of mutiny, he had him thrown into prison in Brazil.

The *Roebuck* did reach Australia eventually, in August 1699, after sailing more than 6,800 miles (11,000 km) across the Atlantic and the Indian Ocean. Dampier anchored in the Bay of Sharks on the arid western coast and collected many plants. He was fascinated by the strange flora of Australia, its trees and fragrant, flowering shrubs, and the abundance of blue flowers. Those intense, luminous blue flowers endemic to Australia would later be classified in the genus *Dampiera*.

Place this P. 96
F. 2.   P. 123.
This very much resembles the Guarauna, described, and figured by Piso.
F. 1.
The Pintado Bird P. 96.

Engraving of sea birds, from A Voyage to New Holland, *in the year 1699: wherein are described the Canary-Islands, Vol. III, by Captain William Dampier, 1703.*

Searching for fresh water, Dampier decided to go north along the coast. If he had turned southward, he might have been the first European to sail around the Australian continent. His choice did lead to making precise maps of the coast of New Guinea and exploring the then-little-known Bismarck archipelago. Despite the difficulties of shipboard life, typhoons and the problems afflicting the *Roebuck*, Dampier never stopped observing the lands and people, winds and currents, and plants and animals they encountered. He drew and took detailed notes, keeping his papers inside a large piece of bamboo sealed with wax to be safe from the seawater.

That was a wise precaution, because the *Roebuck* sank in the Atlantic Ocean. The crew was able to take refuge on some rocks near Ascension Island and Dampier saved his herbarium. After his miraculous return to England in 1710, Dampier was court-martialled for mistreatment of his lieutenant whom he had abandoned in Brazil. Dismissed from the Royal Navy, he returned to his outlaw life and captured one of the Manila galleons and its fabulous cargo of Chinese silk. For the next century, tales of Dampier's exploits would lure many of his compatriots to adventure in the South Seas, including one Captain Cook.

A LIFE LIKE A NOVEL

The adventures of William Dampier, who survived a number of shipwrecks and described far-away lands, were the inspiration for Daniel Defoe's *Robinson Crusoe* and Jonathan Swift's *Gulliver's Travels*.

DAMPIER IN THE GALAPAGOS

More than a century and a half before Charles Darwin, Dampier spent 2 weeks in June 1684 exploring several islands in the Galapagos. He described the enormous tortoises that live on the islands and candelabra cactuses, "as big as a man's leg."

"About a league to the eastward of that point we anchored January the 5th 1688, two miles from the shore in 29 fathom, good hard sand and clean ground. New Holland is a very large tract of land. It is not yet determined whether it is an island or a main continent; but I am certain that it joins neither to Asia, Africa, nor America."

William Dampier

FACING PAGE

HERBARIUM PLATE

Dampiera purpurea R. Br. or *Dampiera brownii*

This specimen was harvested by Joseph Banks in 1803.

Dampiera purpurea R. Br.

see Rajput & Carolin, Telopea 3 : 199
A. S. George II Feb 20. 0.5.
DET

D. Purpurea, Flu.
4 Feb. R. Brown: Can. Pl. I. I. Bennett
2/00. 9

Dampiera pur-
purea
Banks & Mulgrave
1803

THE BEAUTIFUL PLANTS OF THE LEVANT

Tournefort on the Persian border

Joseph Pitton de Tournefort was born in 1656 to a well-off family in Aix-en-Provence. "As soon as he saw plants, he knew he was a botanist," Fontenelle said of him. Leaving school behind, he passionately hunted plants, even in private gardens. He became a doctor and hunted plants in the Alps, Languedoc, Catalonia and the Pyrenees. When he was a professor of botany at the Jardin royal des plantes, he was chosen by the Count of Pontchartrain, Louis XIV's chancellor, to go on a scholarly expedition to the Levant. His story of the journey, in the form of 22 letters to Pontchartrain, did not appear until after he died, having been run down by a carriage in a Paris street in 1708.

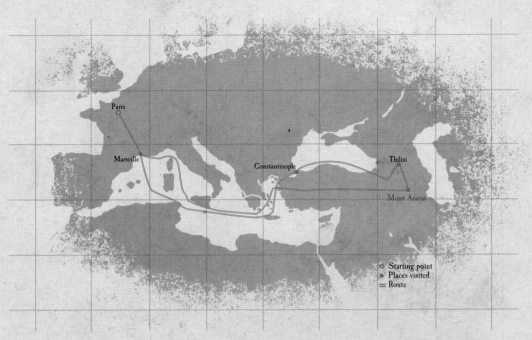

○ Starting point
● Places visited
═ Route

"My Lord, we can no longer defer giving an account of the observations we have made in our walks through the terrestrial paradise," wrote an excited Pitton de Tournefort on July 27, 1701, from the foot of Mount Ararat in Turkey. But before he reached the Holy Land of botany, he had faced many trials.

He left Marseille in April 1700 with Doctor Gundelsheimer and Aubriet, the painter. They spent a year sailing between Candia, Crete, where he climbed the disappointing Mount Ida, "a big ugly, bald hump," and the Greek islands that were then ruled by Turkey. His account of these adventures was lively and full of humor with respect to the discomfort of an explorer's travel. The botanist stayed with the local people after chasing away the "flies, mosquitoes, bedbugs, fleas and ants," ate sea slug soup, and rode "fiery horses" along mountain precipices. He narrowly escaped the "scoundrels of the islands" who threw their victims into the sea with a stone tied around their necks, faced fierce Mediterranean storms, and was harassed and held for ransom by the Turks who accused him of spying and would not believe someone would come so far to "pick up hay." In Constantinople, Tournefort carried a pistol in his pocket and was terrified of the plague, but went into rapture over the buttercups, tuberoses, hyacinths, narcissus and lilies, and dined happily on the melons and cucumbers that grew in the palace gardens.

The small expedition joined the pasha's fleet going through the Bosporus toward Trabzon. Tournefort collected plants on the shores of the Black Sea, all the while railing against geographers who made imperfect maps. Then the three explorers joined a caravan for protection from robbers on their way to Erzurum. Despite the jibes of the merchants, Tournefort constantly got off his horse to "crawl around on all fours, foraging" in the vegetation. After a tense encounter with Kurdish raiders, the scholars arrived at the source of the Euphrates, and chose "one of the prettiest greenswards to spread our cloth upon, in order to refresh ourselves with some of the monastery wine." Finally, after weeks of walking, they reached the territory of the King of Persia in 1701. Their ascent of Mount Ararat was epic, threatened by tigers, avalanches and thirst. Although "the love of plants would carry them through all other difficulties," the scholars thought "they might merit the title of Martyrs to Botany." Their long and perilous journey home took them through Cappadocia and Greece, and they reached Marseille at last, in June 1702.

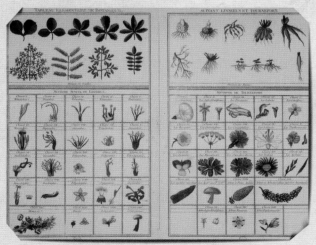

A botanical chart comparing Tournefort's system, based on the flower and other organs (on the right) with Linnaeus's on the left. This engraving once belonged to the Jussieu family.

"We were consoled by the hope that we could visit every corner of this island despite rain, hail, thunder and lightning that were terrifying. We collected plants with our heads down and hoods up, and did not return until evening, loaded with beautiful plants."

Joseph Pitton de Tournefort

FACING PAGE

HERBARIUM PLATE

Vesicaria reticulata

This specimen was collected by Joseph Pitton de Tournefort himself. Tournefort brought back "an infinite number of beautiful plants" to the Jardin du roi, including 1,356 new species and a herbarium with some 8,000 specimens.

Lam. ill. tab. 556. f. 2 =
Vesicaria reticulata. Lam.

FAC. SC. MONTPELLIER.
HERBIER
CAMBESSEDES.
1863.

Vesicaria reticulata Lam.

Vesicaria orientalis,
folio dentatis. Coroll. 49.

Herbier de Tournefort

LICHENS AND FLOWERS
FOR THE FAR NORTH

Carl Linnaeus in Lapland

Carl Linnaeus was born in 1707 into a Swedish family with Lutheran ministers and showed his intelligence and consuming curiosity about plants at a young age. At 5, he already had his own garden plot to care for. At 25, he convinced his father to permit him to study medicine, which may have been a pretext for studying botany. He impressed his professors and met people in Holland, France and England. After a 6-month trip to Lapland, he carried out his great work from his own office. Called "a second Adam," he put the messy nature of Creation in order by classifying and naming the plants, animals and minerals. He died in 1778, having revolutionized botany and the human perspective on natural order.

o Starting point
= Route

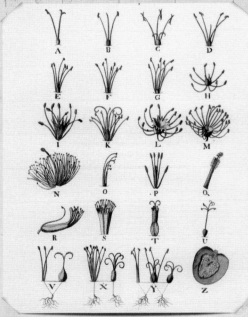

Flower engraving from Systema naturæ sistens regna tria naturæ, in classes et ordines, genera et species redacta tabulisque æneis illustrata, *Carl Linnaeus, 1748*

"With that I left the city of Uppsala on May 12, 1732, which was a Friday, at 11 o'clock, when I was only half a day short of being 25 years old." The young man was jubilant: he was going to spend the bright, brief northern summer exploring Lapland. He had often struggled to make ends meet, but now he had received a fine bursary from the Swedish Academy of Science as a reward for his brilliant studies. For his journey he had a leather haversack containing "one shirt, two pairs of cuffs, two collars, one inkwell, one pen case, a microscope, a magnifying glass, violet to ward off insects, a comb, [his] journal, papers for pressing plants, a guidebook for ornithology, and one for flora."

Thus equipped, the future "Prince of Botany" traveled 1,550 miles (2,500 km) in 6 months, well beyond the Arctic Circle. He went on foot, on horseback and in boats over the region's rivers and swamps. Always astonished, he survived hardship because of his faith and his intense curiosity. He crawled through swamps and icy water up to his belly, fell off a horse, and almost suffocated in clouds of mosquitoes "as numerous as atoms."

He nearly starved to death on Lapland's rough diet of reindeer meat and milk with dried fish full of bugs. But with his eyes wide open, he saw the landscape, the minerals, the animals, the trees and the flowers, and marveled at the beauty of Creation: "The whole world is full of the glory of God."

The farther north Linnaeus went, the less vegetation he found. Forests of alder, birch and poplar gave way to willows, spruce, pine, cloudberries and bearberries. He clambered into grottos and there, after acrobatic searching, found mosses, lichens and ferns. The northern latitudes produced a fleeting abundance of anemones, bellflowers, wild roses, violets "as perfectly white as snow," buttercups that "covered pastures with its brilliant yellow flower," rare coralroot orchids, heather, azaleas and saxifrage. The young botanist turned westward and scaled the Lapland Alps in ice storms and suffered greatly. Still he found new plants in such profusion that he wrote, "I was drowning in wonder, thinking I had found more of them than I knew what to do with." And yet his mind was already organizing, classifying and comparing. Linnaeus returned to Uppsala on October 10, 1732, exhausted and thin but determined to make Nature's exuberance more orderly.

A "LEWD SYSTEM"

Linnaeus, the 28-year-old son of a pastor, scandalized society with his classification of flowering plants according to their sex organs—stamens (male) or pistils (female)—and his lyrical descriptions of astounding sexuality, such as that of poppies, where "20 or more males are in the same female's bed."

FRIEND OF THE LAPLANDERS

Linnaeus survived in the northern lands because of the Lapps (now known as Sami). He admired their nomadic life, their reindeer herding "without war or agriculture, like the Ancients" and their ingenious use of lichens and bark to nourish their herds and provide medicine and clothing.

June 24

"Blessed be the Lord for the beauty of summer and spring, for all that here is more perfect than anywhere else in the world—the air, the water, the green of the plants and the song of birds. I went out this morning to do some botany."

Carl von Linné

FACING PAGE

HERBARIUM PLATE

Linnea borealis L.

These specimens of twinflower were collected in Lapland and Denmark between 1863 and 1868. Linnaeus adopted this small Lapland flower as his emblem. During his journey, having not yet invented his new system of nomenclature with genus and species, he called it "Bellflower with leaves of thyme."

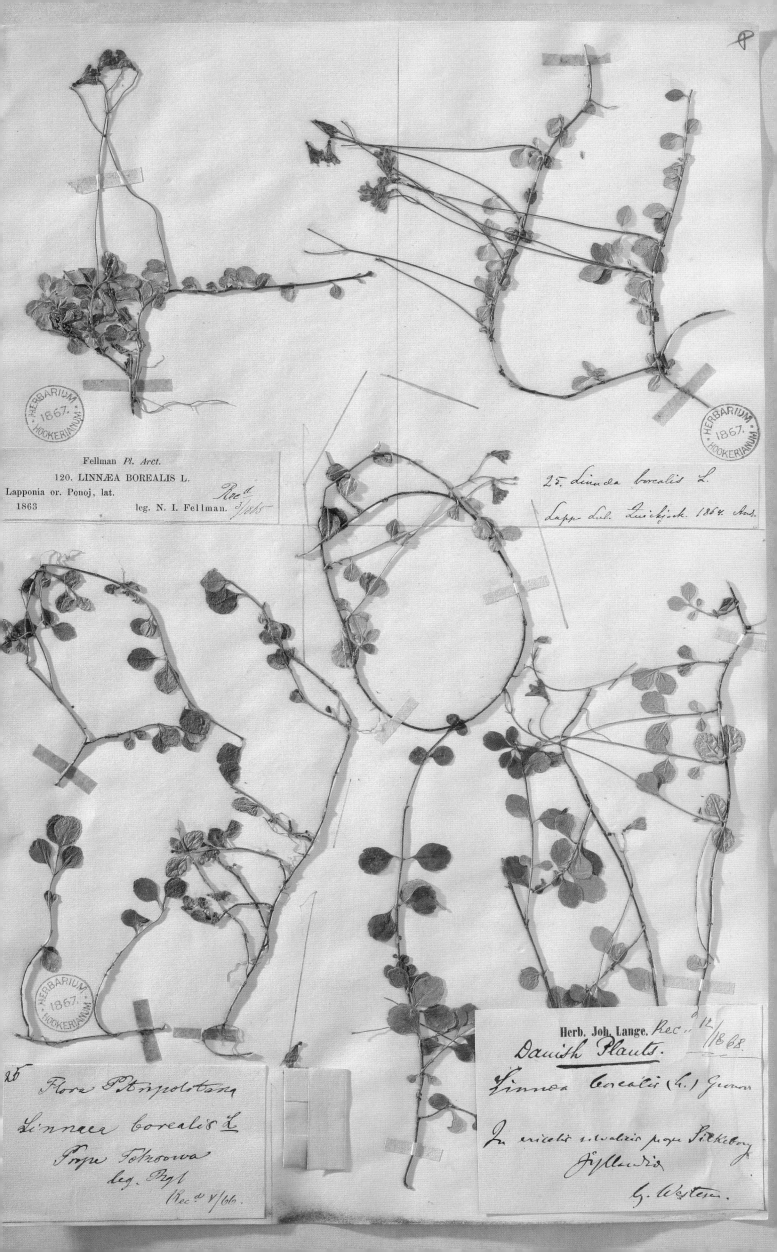

Fellman *Pl. Arct.*
120. LINNÆA BOREALIS L.
Lapponia or. Ponoj, lat.
1863   leg. N. I. Fellman.   Rec⁴ 3/1865

25. *Linnæa borealis L.*
Lapp. Lul. Quickjock. 1864. Ans.

25. *Flora Petropolitana*
Linnæa borealis L
Prope Petrosova
leg. Ryl
Rec⁴ 4/66.

Herb. Joh. Lange. *Rec. 12/1868*
Danish Plants.
Linnæa borealis (L.) Gronov.

In ericetis silvaticis prope Silkeborg
Jylland
G. Westen.

LOST SPECIMENS FROM THE ANDES

Joseph de Jussieu, the unlucky humanist

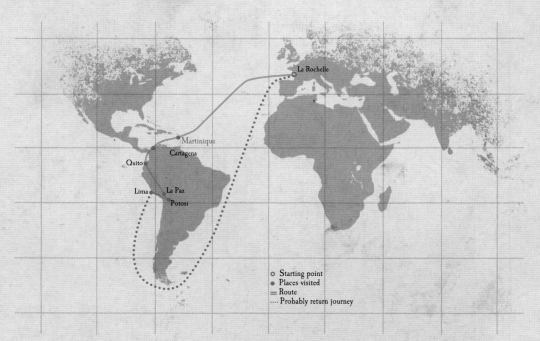

There were three Jussieu brothers. Of the three generations of naturalists in his family, only Joseph de Jussieu was a great traveler. Born in 1704, he was the younger brother of Antoine and Bernard de Jussieu, both eminent members of the Academy and professors at the Jardin du roi. After his studies, Joseph practiced both medicine and botany at the Paris faculty. His brothers thought he tended to fantasy and depression and did not "add any honor to the family." They encouraged him to go on the scientific expedition to Peru led by Charles Marie de La Condamine. Intending to be away for a few months, he ended up spending 36 years in the Andes and returned to Paris mentally and physically broken. He died there in 1779.

○ Starting point
● Places visited
═ Route
···· Probably return journey

A debate was raging in 1735: was the Earth flat at the poles like an orange or pointed like a lemon? In order to settle the argument, Louis XIV financed two expeditions to measure the curve of a meridian, one to Lapland and one to Peru. Joseph de Jussieu embarked for Central America with the chemist and geographer Charles Marie de La Condamine and five other scholars. After long stops in Martinique and Panama, where Jussieu happily indulged in botanical studies, the expedition reached Lima. The scholars began the superhuman task of calculating the circumference of the Earth by triangulation. The harsh climate of the Andes, resistance from the indigenous people, suspicion from the Spanish authorities, and especially internal disputes and the shortage of funds dragged their work out from year to year. In contrast, the Lapland part of the expedition had been a success and the king threatened to cut off support for Peru. But sending letters back and forth to Paris took 2 years, and La Condamine used the delay to advance his calculations.

Meanwhile, Jussieu was exploring the jungle and the summits of the Andes, making notes, discovering the effects of coca, collecting scented heliotrope, white cinnamon, nasturtium and periwinkle. Then he was needed to fight a smallpox epidemic and devoted himself to helping the poor. Seeing many tropical fevers, he became interested in cinchona. He prepared an extract of the plant and wrote up his complete findings in 1737. He gave this dossier to La Condamine who sent them to the Academy, but the Academy attributed the work to the geographer and not the botanist. Unfortunately, that was the only part of Jussieu's work that ever arrived in France.

The botanist went deep into the Amazon forest in a dugout canoe. Soon he abandoned his rifle and was accepted by the Indians. He lived in their villages

Drawing of cinchona by Joseph de Jussieu

and studied insects and plants. But when he returned to Quito he was a witness to the beheading of the last Inca, symbolizing the ruin of Indian dignity. Jussieu sank into a depression. He was not able to return to France despite his brothers' entreaties and La Condamine's departure in 1744. In 1746 he learned that his mother and brother had died and he devoted himself to taking care of the Indians, first those injured in the great earthquake in Lima, and then the forced laborers in the silver mines at Potosi.

In 1771, friends put him on a ship quite forcefully, along with his papers and the collections that had not been lost. But his baggage never reached its destination. After 36 years of hard work, Joseph de Jussieu landed in Europe, his mind and body broken and his hands empty.

HIGH-ALTITUDE TRIANGULATION

In order to measure the Peruvian meridian, the scholars established triangles whose angles they measured from high places. But at 13,100 feet (4,000 m) the mules died of cold and the llamas refused to move. The expedition did succeed, however, in calculating one degree of meridian with an error of only 72 feet (22 m).

THE WEEPING TREE

When returning to Europe, La Condamine chose to descend the Amazon in a dugout canoe to reach Cayenne. He did some botanical observation while waiting for a ship and discovered the rubber tree and its stretchy, impermeable sap.

FACING PAGE

HERBARIUM PLATE

Cinchona officinalis

This specimen of cinchona was harvested in South America. Joseph de Jussieu wrote a memorandum on all the known properties of the various species of cinchona. Linnaeus named the genus in 1742 after the Second Countess of Chinchón. According to legend, she was cured of malaria by a botanical remedy made of the powdered bark of a native tree.

"With the help of two local Indians whom I had taken to help me, I could only find eight or nine young cinchona plants suitable for transporting that day. I had them put into a good-sized crate with earth from the same place."

Charles de La Condamine

C. officinalis (167)

Strong growing var / Bonplandiana?
Nilgiri Cinchona Plant.
Can Mur Petcock 19/18

167

Same as 168,

IN THE EMPEROR'S GARDENS

D'Incarville in China

Pierre Noël Le Chéron d'Incarville was born in 1706 into a family of minor nobility at Louviers and studied at the Jesuit seminary in Rouen. At 21 he entered the novitiate of the Company of Jesuits and was sent to teach in Quebec from 1730 to 1739. Fascinated by botany and silkworm culture, and also a glassmaker, he went to China in 1741, settling in Beijing. Having written to Bernard de Jussieu, he was named a correspondent of the Academy of Science in Paris. In 1752 he became the botanist and manager of China's imperial court gardens. He died of a fever in Beijing in 1757, at the age of 50.

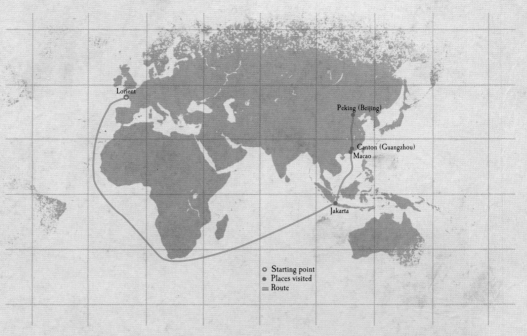

Lorient

Peking (Beijing)

Canton (Guangzhou)
Macao

Jakarta

○ Starting point
● Places visited
═ Route

"We left L'orient, the Indies company's port, on January 19, 1740," wrote Pierre d'Incarville. A year later, the young Jesuit landed in Canton. There, he studied Chinese and then made his way across the empire to Beijing. During that year, Pope Benedict XIV had condemned Confucian religion and the English and the Dutch were putting more pressure on China. The Middle Kingdom did not appreciate any of this. Hostility toward westerners was growing, especially toward missionaries. Prudently, the Jesuit fathers did not go out much, but d'Incarville was busy. He observed the plants in the gardens of his residence and quickly sent a herbarium to his friend Bernard de Jussieu, the conservator of the Jardin du roi. In an attic, he found a copy of a medical text from 1505 that contained more than 400 exquisite drawings of flowers.

When tensions were lessened, d'Incarville traveled around China. He gathered seeds of the kiwi, tree of heaven (*Ailanthus*), golden rain tree and Japanese pagoda tree, which he sent to France, England and Russia. Some colleagues were jealous of him, but he was recognized by the Academy of Science, which named him an official correspondent in 1750.

D'Incarville's greatest desire, however, was to enter the gardens of the Forbidden City. Looking for a way to stimulate the Emperor's curiosity, he wrote to Jussieu, "If you could send me some seeds and bulbs that I could present to the Emperor, I could become known to him, at least, as someone curious about flowers and later as a botanist, which would give me an opportunity to see many plants that I would otherwise probably never see." Jussieu sent him sent him many plants — tulips, wallflowers, carnations, lily-of-the-valley, rosemary, fruit trees — but it was mimosa, also known as sensitive plant, that opened the gates of the palace to the Jesuit botanist. In 1753, he offered

two roots of this tree to Emperor Qianlong, who was immediately taken with this tree that seemed to have a soul; its leaves closed when they were touched. The mimosas were planted in the royal greenhouses and d'Incarville was "ordered to go see them frequently." With the Emperor's support, his research increased, covering silk, varnish, drugs and plants. He sent valuable memoranda to his international correspondents, along with the seeds of trees and shrubs identified as *arbor incognita sinarum*, unknown Chinese tree, until he died in 1757.

Plate illustrating Chinese flowers, sent by d'Incarville to Bernard de Jussieu in 1742. This plate is one of several copies d'Incarville had made in Beijing from a manuscript kept in the imperial library.

STINK TREE

Among the seeds sent by d'Incarville to Jussieu there were some of *Ailanthus altissima*, the Tree of Heaven, which was cultivated to feed silkworms. Despite their unpleasant odor, the trees became widespread as ornamentals in Europe and North America.

SEEDS OF LIFE

It was very difficult to keep seeds viable during the year-long sea voyage from China to Europe, so d'Incarville often used the services of correspondents in Saint Petersburg, reached by Beijing–to–Moscow caravans.

"The Emperor loves flowers; he had rooms expressly built that give him a view of a small hill covered with chrysanthemums, which bloom in an astonishing variety of colors."

Pierre Noël Le Chéron d'Incarville

FACING PAGE

HERBARIUM PLATE

Incarvillea forrestii

This specimen was harvested by George Forrest in western China.

Determ. I. Vassilczenko.

Incarvillea Forrestii
Fletch.

30.V
195ᵭ   Опр. И. Т. Васильченко

YUNNAN, WEST CHINA.
Coll. GEORGE FORREST.   No. 21526
EX HERB. HORT. REG. BOT. EDIN
                                      19

Alt.
Locality *Forrestii R. Uckhu*
21526. **Incarvillea grandiflora**, Franch. var. vel aff. Plant of
1½–2 ft. Flowers pale rose. Openings in thickets
and amongst scrub on alpine meadows on the Chien-
chuan-Mekong divide. Lat. 26° 40′ N. Long. 99°
40′ E. Alt. 11–12,000 ft. July 1922. N.W. Yunnan.

BREAKING THE SPICE MONOPOLY

Pierre Poivre in the East Indies

Pierre Poivre was born in 1719 to a silk-making family in Lyon. After studying theology, botany and painting, he entered a missionary order. At 22, he was sent to China and Cochinchina, where he was fascinated by everything. On the return journey, his ship was attacked and he lost his right hand to cannon fire; the right hand is used to give blessings. Imprisoned in Batavia (Jakarta), he gave up the priesthood and devoted himself to the Asian spices he was trying to acclimatize to the French colonies. From 1748 to 1771 he made five expeditions, three for the French East India Company. He was named Intendant of Mauritius and Réunion, and died in Lyon in 1786.

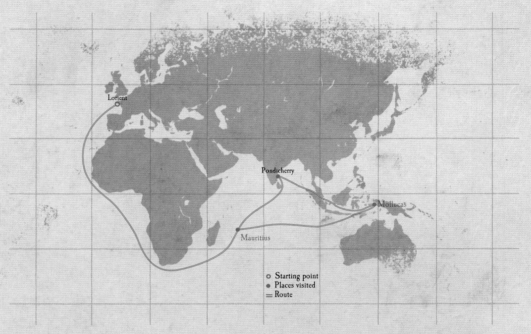

Lorient
Pondicherry
Moluccas
Mauritius

○ Starting point
● Places visited
＝ Route

Pierre Poivre was sent to Asia as a young missionary in 1741. Fascinated by the Far East and Chinese civilization, he took great interest in rice paddies, lacquers and bamboo, but did not convert anyone to Christianity. His unhappy superiors recalled him to France, and on the voyage his ship was attacked by the English and a cannon ball removed his right hand. Poivre spent many months in Jakarta, wounded and imprisoned. He observed the monopoly on nutmeg and cloves that the Dutch were enforcing by terror in the Moluccas. Poivre discovered their secrets and passed them to the French East India Company. "I learned that the Dutch had little strength in the Moluccas, where they acquire their two fine spices just from the two islands of Ambon and Banda. Ambon alone produces the entire quantity of cloves that the Dutch need to supply the stores of every nation, and the little Island of Poulo-ai, only about two leagues in circumference, produces enough nutmeg for the entire universe to consume." He added that these spices also grew on other islands. All it would take would be some determined men and a good ship, in order to take some plants and acclimatize them to "our French islands" (Mauritius and Réunion) and thus acquire a share in this highly lucrative trade.

From then on, Poivre devoted himself to this mission for the glory of his country. The Company agreed to support him but did not provide money, nor a ship and crew. With outstanding ingenuity and self-sacrifice, Poivre organized two expeditions in 1749 and 1750. He was able to buy nutmeg seed on the black market and germinated 32 plants, but only five of them were living when they reached the French islands. And worse, the colony's pharmacist poisoned these precious plants "by covering the roots of each one with a cartload of cold, sterile, mucky clay." Poivre suspected

him of being in the pay of a new director of the Company, who was of Dutch origin.

Stubbornly, Pierre Poivre set up a third expedition in 1754, which was unsuccessful because he was given a "bad frigate." Having become the official intendant of the French islands, he launched a fourth expedition that brought back nothing but plants that had died of drought. His fifth attempt, in 1751, finally was successful but the plants did not thrive as much as he had hoped. It was not until 1780, when they were planted in the hot, humid climate of French Guiana and the Caribbean that French spices finally flourished.

PAMPLEMOUSSE BOTANICAL GARDEN

Pierre Poivre was named Intendant of Mauritius by Louis XIV in 1767, and he acquired land there, where he acclimatized the precious spices and distributed them to Madagascar and the Antilles. In 1768 he was visited by Joseph-Philibert Commerson, the botanist of Bougainville's expedition. They worked together for 2 years, designing and planting a botanical garden in the Pamplemousse district. They added many plants to their collection including Antilles laurel, cacao, mango, new cocoyam, sago palm, breadfruit, cinnamon and sugar cane from Java.

TRAVELS

OF A

PHILOSOPHER:

OR,

OBSERVATIONS

ON THE

MANNERS AND ARTS

OF

VARIOUS NATIONS

IN

AFRICA AND ASIA.

TRANSLATED
FROM THE FRENCH OF M. LE POIVRE,
LATE ENVOY TO THE KING OF COCHIN-
CHINA, AND NOW INTENDANT OF THE
ISLES OF BOURBON AND MAURITIUS,

Title page of the 1770 English translation of Pierre Poivre's Voyages d'un philosophe, *1769*

"I thought that the only way to acquire clove plants was to use some local vessel to go myself, or send someone I trusted, to seek those plants on the islands where they grow but where the Dutch had not noticed them."

Pierre Poivre

FACING PAGE

HERBARIUM PLATE

Myristica fragrans

This specimen of a nutmeg plant was collected in the Moluccas in 1899.

141

Kew

UNDER THE BAOBAB TREES

Michel Adanson in Senegal

Michel Adanson was born in Aix-en-Provence in 1727, into a family with Scottish roots. A brilliant student, he eagerly studied botany with the "illustrious members of the Academy, Bernard de Jussieu and René Antoine de Réamur, at the Jardin du roi. At the age of 22, Adanson left for 4 years in Senegal and came home with a monumental collection of specimens and notes, for which he was named to the Académie royale and the Royal Society in London. He published *Familles des plantes*, an innovative classification of plants into 58 families. In the end, his research so obsessed him that he left his wife and children and devoted himself to creating an impossible universal encyclopedia of natural science, on which he worked until his death in 1806.

Brest
Lorient

Tenerife

Saint-Louis

○ Starting point
● Places visited
≡ Route

In March 1749, a 22-year-old botanist, serving as a clerk with the French Indies Company, took ship at Lorient. His investment was small but his project was ambitious. Michel Adanson was going to explore "equatorial Africa that had never been visited by a naturalist" and hoped to find "a vast field of observations to be harvested." He was enthusiastic about the voyage: the "beautiful, calm" sea, thousands of porpoises, phosphorescence in the ocean, the stop at Tenerife and so on.

Arriving in Senegal, he was provided with "a canoe, some natives and an interpreter" to go and reconnoiter the country's products. His curious mind was enthralled by the "sky, climate, inhabitants, animals, lands and plants." As an independent explorer, steeped in humanism and Rousseau's myth of the "noble savage," he learned the Wolof language and admired the customs of the indigenous people. "The pastoral setting, with the huts in among the trees, the idleness and indolence of the natives lying in the shade ... all made [him] think of the first men, and [he] felt as if [he] I were seeing the world at the time of its birth."

Adanson borrowed a dugout canoe and thoroughly explored the banks of the Niger River, with myriad islands, sandbanks, mangroves and forests. As he hunted animals and plants, he described the mangroves that "grow only on the river banks where the seawater rises twice a day," tamarind trees, huge bombax trees, many varieties of willow, palm, sensitive plant and thorny acacia. Gum acacia, from which gum arabic was produced, interested him both as a botanist and as a representative of the Company. But the encounter of his life was with a truly enormous tree that "looked less like a single tree than like a whole forest." Adanson was enthralled by the baobab (also called monkey-bread tree), its trunk that took

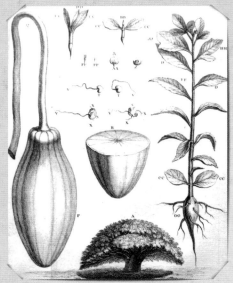

Engraving of a baobab from Adanson's Mémoires de l'Académie royale, *describing a new genus of tree, the baobab, observed in Senegal, 1765.*

him 13 times stretching his arms out to go around, its branches that each "would have been a monstrous tree in Europe." He devoted a whole book to the baobab, calculating that the oldest specimen went back to the Biblical flood, and drinking its mucilaginous innards to fight malaria. For after 4 years, fevers and seasickness had ruined the young man's health; he had to return to France.

Before his return, Adanson had sent the Jardin du roi thousands of specimens of shells, dried plants and seeds. But he did not bring any living specimens home because a storm pursued him all the way from the Azores to Brest, ruining almost 300 tree seedlings. The few that survived perished in the freezing weather of the extreme winter of 1753, before they even reached the royal greenhouses.

GREATNESS AND DOWNFALL OF A BOTANIST

Adanson's *Histoire naturelle du Sénégal*, published in 1757, was a money-pit for the author. In 1764, after 6 years of negotiations, he decided to sell his collections of shells, fish and plant specimens to the Cabinet du roi. His meager royal pension and stipend as a member of the Academy were eaten up by the cost of publication, and then the Revolution took away all his income. His poverty was such that he refused an invitation from the Academy because he had no shoes, and died penniless.

"*I climbed up on his shoulders carrying my rifle, some birds and a package of plants ... Having escaped that dangerous step, I saw a most beautiful floating plant; it was an amaranth with silky, silvery flowers. I forgot everything in a flash and although Banbara was still up to his chin in the water, I attempted to uproot that beautiful flower.*"

Michel Adanson

Juin 1852.

Adansonia digitata L.

Le pédoncule fructifère, long d'un
pied et demi, paroit avoir une
structure très-singulière, que
Walpers a décrite dans Mohl et
Schlecht. Bot. Zeit. 1852, p. 295–99.

« Bei einem Querschnitt bemerkt
« man schon mit blossem Augen und
« auf das Allerdeutlichste, dass derselbe
« entsprechend den pentameren Blüthen-
« und Fruchtverhältnissen, aus fünf
« verschiedenen völlig geschlossenen
« Holzcylindern gebildet wird, deren
« jeder seine eigene Mark-höhle ein-
« schliesst, und ringsherum von einer
« besonderen sehr starken Rindschicht
« umgeben ist! » p. 297.

Flora of West Africa
Det. Hutchinson & Dalziel

Adansonia digitata L.   Cav. Diss. 5. p. 298.
tab. 157)

HERB. J. GAY.
Presented by Dr. Hooker, February 1868.

(Senegal)   Roger dedit Junio 1821.

FROM BOUGAINVILLEA TO HYDRANGEA

Philibert Commerson and Jeanne Barret around the world

Philibert Commerson was born in a village in the Dombes region of Burgundy in 1727. He did not concentrate on his medical studies in Montpellier, because he had only one passion: botany. He sometimes harvested plants for his herbarium from people's gardens without permission. He became a doctor and married a woman he loved very much, but who died when their son was born. Despairing, Commerson moved to Paris. He had the support of the Academy of Science to join Bougainville's 1767 expedition. After hunting plants in South America, Indonesia, Polynesia and Madagascar, his last stop was the Île de France (now Mauritius), where he died in 1773, at 46, never having seen his beloved son again.

Map labels: Rochefort · Tahiti · Rio de Janeiro · MADAGASCAR · Mauritius · Strait of Magellan

○ Starting point
● Places visited
═ Route

In the harbor at Rochefort, in February 1767, a man boarded *L'Étoile*, one of the ships in the first French fleet to circumnavigate the Earth, under the command of Monsieur de Bougainville. That man, Philibert Commerson, was a model Enlightenment scholar: a naturalist with a passion for botany and a keen interest in fish, insects, birds, weather and geography. Still, he was not always very observant, or perhaps he was distracted, because when he was provided with a "valet paid for by the King" by the name of Baré, he did not recognize Jeanne Barret dressed as a man. And yet she was the housekeeper who had been looking after his son and organizing his household and his plant specimens. There was a royal edict forbidding women aboard ship.

At Rio de Janeiro, *L'Étoile* met up with *La Boudeuse*, and while they waited for the southern hemisphere summer, Commerson happily searched for plants. He found some 1,800 species of vegetation, including "a wonderful plant with large, rich violet flowers" that we now know as bougainvillea. As they went through the Strait of Magellan, his enthusiasm continued, plant-hunting with the tireless Baré and helpful Patagonians "who brought us the species they saw we wanted." Then they crossed the Pacific, and made a marvelous stop in Tahiti in April 1768, where they spent 9 days discovering the men and women "born under the most beautiful sky, living on the fruits of a fertile, uncultivated land." Whether the Tahitians were experts in love or in physiognomy, they immediately saw that Baré was a woman. Bougainville was thunderstruck. How

could that be? The valet was a woman? "That already expert botanist whom we had observed following his master on all his plant-hunting explorations, in the snow and icy mountains of the straits of Magellan, even on those rude hikes carrying the food, weapons and plant books with courage and strength, the one nicknamed by the naturalist his 'beast of burden'?"

On Mauritius, Commerson met his old friend Pierre Poivre and accepted his hospitality, while they worked together, always aided by Jeanne, on the famous botanical garden of Pamplemousse. In 4 years, Commerson harvested 1,574 plant species, including hydrangea. On a 6-month journey to Madagascar, which he considered "the promised land for naturalists," he found another 495. But "exhausted from late nights and fatigue" Philibert Commerson died in March 1773. He left Jeanne his treasure, more than 6,000 specimens, that she faithfully sent to the Jardin du roi.

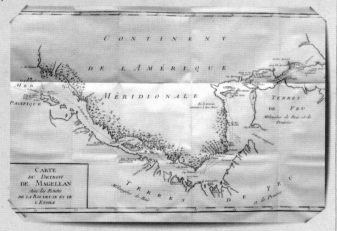

Map of the Strait of Magellan showing the routes taken by La Boudeuse *and* L'Étoile, *from* A Voyage Round the World... in the Years 1766, 1767, 1768 & 1769 ... in the Frigate La Boudeuse and the Store-ship L'Étoile, *L. de Bougainville, 1769*

THE APPLE OF PARADISE

The Europeans were fascinated by the freedom and sensuality they saw in the Tahitian people. Commerson called the island Utopia and Bougainville named it New Cythera, after the birthplace of Venus Aphrodite. The name survives in the *Spondias cytherea* fruit, endemic to Polynesia and known as golden apple in the Caribbean and Wi apple in Hawaii.

A FORGOTTEN TRUNK

Commerson's notes were returned to France along with his plant collections, in 1774. But at the Jardin du roi, Bernard de Jussieu was not interested in the papers and they remained in a trunk in Buffon's attic until they were discovered, years later.

"This plant whose leaves and appearance may deceive is dedicated to the valiant young woman who, taking on the clothes and temperament of a man, had the curiosity and audacity to see the whole world, by land and sea, accompanying us although we were unaware."

Philibert Commerson

La Rose du Japon *acaï-sant*

Flores annoui cærulei

Petale ut in apile longe v anguste tubulosa
hinc quinque v quadribula, lobis orbiculatis ad faucem netro tuti iacitis, seux
acilen occupata squamula 5 clavium poicuimera circumposites, pro
de cætum flamina ex opta tubifiesca non finib excita sint, de Style
nec sibi constitti. Sed stigmata duo acuminata videbantur

hortensia
Lam. Dict.

lam. n.° 57

HARVESTING SOUTHERN HEMISPHERE PLANTS

Banks and Solander in the South Seas

Joseph Banks, son of an English landowner, inherited his father's fortune when he was young. His mother, a very cultivated woman, encouraged him to devote his intelligence and his money to the natural sciences. When he was 23 and a student at Oxford, Joseph Banks joined a scientific expedition to Labrador and, when he returned home, was elected to the Royal Society. At 25, he went on Cook's famous expedition and then on a trip to Iceland, and then spent the rest of his life promoting science in general and botany in particular. Banks became director of the Royal Gardens at Kew in 1773, founded the Royal Horticultural Society in 1804 and sent plant collectors to the four corners of the Earth. He died in 1820.

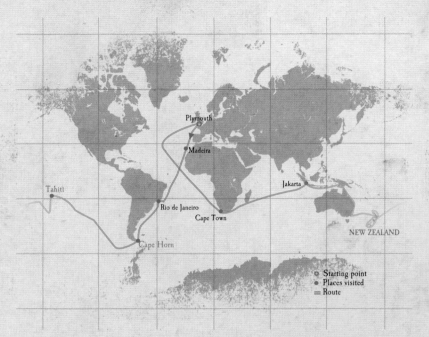

Daniel Solander was born to a Swedish Lutheran family in 1733. He was a brilliant student of botany and Linnaeus's favorite student. Linnaeus sent him to England to preach the virtues of his new system of classification. Solander was welcomed into London's society of scholars and worked on the natural history collections of the British Museum. In 1764 he was admitted to the Royal Society and became close friends with Joseph Banks, who invited him to join Cook's expedition. When he returned, he became conservator of the British Museum and librarian of Kew Gardens, where Banks was director. He refused to participate in Cook's second expedition but faithfully accompanied Banks to the Faroe Islands. He died in London in 1782.

The most fruitful scientific expedition of its time! The most enriching human adventure! The largest plant harvest ever described! There was no shortage of superlatives to describe the epic journey of Captain James Cook and the *Endeavour* in the Pacific Ocean. The primary goal was to observe the transit of Venus between the Earth and the Sun, a rare astronomical event that helped measure the solar system more accurately. Cook also wanted to discover the mysterious Terra Australis that western navigators had been seeking for centuries.

His passenger, Joseph Banks, was determined to take inventory of the natural resources of the lands where they stopped. To do so, the young, wealthy nobleman had hired a team of eight, including his friend the botanist Daniel Solander, a secretary, two painter-naturalists and four plant-hunters. He paid for renovating the ship to suit their purposes, and supplying it with a solid library of natural history, all sorts of devices for capturing insects and fishing in the coral reefs, an underwater telescope, an assortment of bottles of formaldehyde, salts for the conservation of seeds, beeswax, myrtle and so on, all at an estimated cost of £10,000, compared to the purchase price of the *Endeavour*, a mere £2,840.

Ready at last, *Endeavour* left England for the South Seas on August 26, 1768. It was a robust three-masted bark, a former coal-carrier 100 feet (30 m) long, with a 96-man crew. At Madeira, the first port of call, Banks and Solander rushed to disembark and collected 700 specimens in less than a week. A month and a half later, *Endeavour* reached Brazil, where the suspicious Portuguese governor refused to let them land. But the botanists slipped

Map of the coasts of New Zealand explored in 1769 and 1779 by Captain Cook, commander of the Endeavour

out of their cabin through the window and down a rope to a skiff, which they rowed to the shore for nightly explorations. At the next stop, Cape Horn, the scientists were trapped by a snowstorm. Banks tried to save his companions but, despite his encouragement, two exhausted servants refused to go on walking and froze to death.

In April 1769, the *Endeavour* reached Tahiti in time to observe the transit of Venus. In 3 weeks, Banks and Solander scooped up enough flora and fauna to supply the two painter-naturalists, but one of them died and Sydney Parkinson alone was faced with that gigantic task under difficult conditions.

(continued on page 66)

HARVESTING THE OCEAN

Even at sea, Banks and Solander did not stop collecting, and from the deck of the *Endeavour* gathered algae, birds, fish, mollusks, crustaceans and even, off the coast of Brazil, a whole collection of butterflies.

COLONIZING AUSTRALIA

Banks was an ardent proponent of sending British colonists to Australia. With his personal knowledge of the climate and soil, he designed a kit of cereal grains, vegetables, aromatic herbs, fruits and berries that would meet the needs of the pioneer farmers.

"The island of Ohaena is 15 leagues in circumference. The richest imagination could not picture a more enchanting sojourn. The shore is bordered with heavy-laden fruit trees, and between them one can see the coconut palms waving their fertile fronds."

Joseph Banks

FACING PAGE

HERBARIUM PLATE

Scleria lithosperma (formerly *Schoenus filiformis*)

The specimen on the right was collected by Sir Joseph Banks himself.

9

Scleria lithosperma (L) Sw.
Var. lithosperma
= S. filiformis Sw. TYPE.   JOHN E. FAIREY, III   *May 24, 1987*

lithosperma Swartz
Scleria (filiformis Swartz)
C.B. Clarke
Aug. 1890

FLORA JAMAICENSIS.

HERB: BOTANICAL DEPARTMENT, JAMAICA.

Rec.d XII. 1886.

No. **1543**

Scleria filiformis Sw.

Locality: ....

Collector: .... *I. C.* ....

Date ... *Oct* 188 *6*

1,000—19.8.86.

Schoenus filiformis Sir J.B.
—— lithospermus Sp. plant.
Carex lithosperma Murray.

Herbarium of the late BISHOP GOODENOUGH.
Presented by the Corporation of Carlisle, June 1880.

Carex

Sir J.B.

— SWARTZ !

Cook described the painter at work, "Not only did the flies cover his subject, but they ate the paint off the paper as quickly as he applied it."

From Tahiti, *Endeavour* sailed toward New Zealand. Cook devoted 6 months to mapping its contours in detail. But the Maori were combative and the crew could only go ashore rarely and quickly. The naturalists did manage to harvest about 60 new species, among them the kowhai (*Edwardsia tetraptera*) with its spectacular bunches of golden yellow flowers, karaka (*Corynocarpus laevigatus*) with its toxic fruit, and two shore plants that Cook ordered gathered to prevent scurvy: *Lepidum oleraceum* (still called Cook's scurvy grass) and *Apium prostratum*, or sea celery.

Three weeks later they arrived on the east coast of Australia, which had never yet been explored. That was the beginning of a great voyage of discovery and the largest harvest of unknown plants ever carried out. On April 28, 1769, *Endeavour* dropped anchor in a large, tree-lined bay that earned the name Botany Bay. For a week, Banks and Solander gathered plants in a frenzy, from the swamps to the shore to the great primeval forests: eucalyptus, acacia, mimosa, passion flowers, morning glories and orchids ... all eventually acclimatized in Europe. Parkinson, the painter, worked late into the night by candlelight but could not keep up. He had to settle for sketches with notes indicating colors, while the specimens rolled in damp cloth piled up around him. The next 4 months were a nightmare: the *Endeavour* was trapped inside the Great Barrier Reef, 1,240 miles (2,000 km)

Drawing of Canavalia rosea *by Sydney Parkinson on the voyage of* Endeavour *under the command of Captain Cook*

of shallow reefs and savage currents. It took all of Cook's navigational expertise to get them out of that "crazy labyrinth" and steer a course for Java. They found no rest there, for malaria, dysentery and tuberculosis carried off the weakened sailors. By the time they reached the Cape of Good Hope, 34 men were dead, including the astronomer, Charles Green, and the talented painter, Sydney Parkinson. Banks and Solander were gravely ill, but they lingered near Cape Town to do some collecting, which was meager despite South Africa's botanic wealth.

Glory awaited them in England when *Endeavour* landed on July 12, 1771. The expedition's results were prodigious: 30,000 plant specimens — more than 1,400 of them new to science — and 1,000 animal species. From then on, scientists were entitled to a place in all voyages of discovery.

BREAD AND MUTINEERS

In 1769, Cook's botanists discovered the breadfruit tree in Tahiti. Sydney Parkinson describes it precisely in his journal: "This tree, which yields the bread-fruit so often mentioned by the voyagers to the South-seas, may justly be stiled the Staff-of-life to these islanders; for from it they draw most of their support." Joseph Banks saw it as the ideal food for the slaves on British plantations in the Caribbean, and so he asked that a 1788 expedition gather some plants: the expedition led by William Bligh on the *Bounty*.

"The manner in which the New Zealand Warriors defy their Enemies," from
A Journal of a Voyage to the South Seas, in His Majesty's Ship, the Endeavour *by Sydney Parkinson, 1773*

"Our plant harvests are now so immense that it was necessary to take extraordinary care so that they did not rot in the herbarium. Thus, today I brought all the specimen sheets, nearly 200 of them, to the land where I spread them out on a sail all day, turning them frequently."

Joseph Banks

Panicum polygonum sp. nova ?

Sir J. Banks.

maximum Jacq.

SEARCHING FOR SOUTHERN PLANTS

Forster and Son in the South Pacific

Johann Reinhold Forster was born in Prussia in 1729. Having become a professor of languages and natural science, he moved to England in 1766. Since he was well-known for his natural science work and for his English translation of the story of Bougainville's expedition, Johann Forster was chosen to replace Banks on Cook's second voyage to the south seas. He brought along his son, Georg Adam, age 18, as a draftsman. Young Forster was a talented ethnographer and his detailed account of the voyage was highly successful. He continued his career as a traveling writer, and also engaged in politics with German revolutionaries.

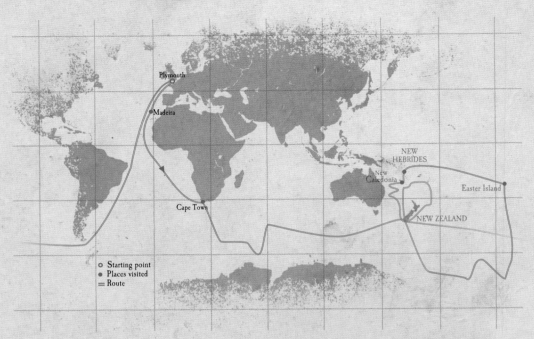

○ Starting point
● Places visited
= Route

One year after his triumphal return to England, the tireless James Cook took to the sea again, with a grandiose goal: "to go south, farther than any man has ever gone." But neither Joseph Banks, who had sponsored the first voyage around the world, nor the botanist Solander was going. Banks had such demanding requirements for equipment and staff that Cook feared for the safety of his ship. Taking their places were Johann Reinhold Forster, an emeritus professor of natural science with a somewhat difficult temperament, and his son Georg as the artist.

The expedition's two ships, *Resolution* and *Adventure*, left Plymouth on July 13, 1772, headed for Madeira and the Cape of Good Hope, where they stayed for several weeks. From there, they made their first forays into the Antarctic high latitudes. "The icefields appeared, in different parts of the horizon about us on the 15th in the morning," wrote Georg Forster. "The weather, which was already foggy, became thicker towards noon, and made our situation, amidst a great number of floating rocks of ice, extremely dangerous." The *Resolution* managed to cross the Antarctic Circle and reach as far as 71° south, dodging giant icebergs before finally being stopped by pack ice. The crew and scientists were plagued by rheumatism because of the glacial cold. Their diet of rotten bread gave them scurvy with its "excruciating pains, livid blotches, rotten gums, and swelled legs."

Seeking provisions and repairs to the *Resolution*, Cook

sailed toward New Zealand, Tahiti, the New Hebrides, Easter Island and New Caledonia, which he discovered in passing. The naturalists rushed onto each land, observed the natives, collected plants, hunted and painted birds. On their incredible 3-year voyage, covering 70,000 miles (113,000 km), they had only 290 days at anchor, yet their contribution to science was considerable. They filled in many of the gaps left by Banks and Solander concerning the southern flora, and discovered many new species, in particular the large trees such as the *Araucaria* (Norfolk pine and monkey-puzzle, for instance), mahogany, Antarctic beech and Chilean fire tree (*Embothrium coccineum*).

Their welcome back in England was not what their exploits deserved: the government refused to pay Johann Forster the £1,000 he had been promised and young Georg had to sell his paintings at auction. Worse yet for a botanist, Forster's new plant descriptions were not published until 1844, and in the meantime were attributed to others.

COOK'S PINE

In 1774 at the most southerly point in New Caledonia, James Cook sailed past an island covered with trees like tall green columns. They were *Araucaria columnaris*, conifers endemic to that archipelago. He did not set foot on the island, but the species is known as Cook's pine and the island as the Isle of Pines.

FAUNA OF THE ICE

The first scientists, and perhaps the first humans, to cross the Antarctic Circle, the Forsters discovered the glacial ecosystem and its unique fauna, including snow petrels, emperor and king penguins and fur seals.

Illustration of one of the ships and a boat from James Cook's second expedition among the icebergs. From Voyage dans l'Hémisphère austral, et autour du monde, fait sur les vaisseaux de Roi, l'Aventure, & la Résolution, en 1772, 1773, 1774 & 1775. *James Cook*

"... from the mast-head (we) observed one hundred and sixty-eight ice islands, some of which were half a mile long, and one less than the hull of the ship. The whole scene looked like the wrecks of a shattered world, or as the poets describe some regions of hell ..."

Johann Reinhold Forster

FACING PAGE

HERBARIUM PLATE

Semecarpus atrum

This specimen of cashew or tar tree was collected in New Caledonia by the Forsters themselves.

2

FLOWERS OF THE FORBIDDEN EMPIRE

Carl Thunberg, botanist and smuggler, in Japan

Carl Peter Thunberg was born in Sweden in 1743 and studied with Linnaeus. He studied medicine and natural history in Uppsala, then in Paris. In 1772 he went to South Africa to improve his spoken Dutch so that he could pass as an employee of the Dutch East Indies Company and thus enter Japan. He stayed in Japan for a year, in 1775–76, then traveled to Ceylon. Returning to Amsterdam in 1778, he published his botanical research under the title *Flora Japonica* in 1778, and later his account of the journey. He was nicknamed the "Japanese Linnaeus," and published more than 100 works before he died in 1828.

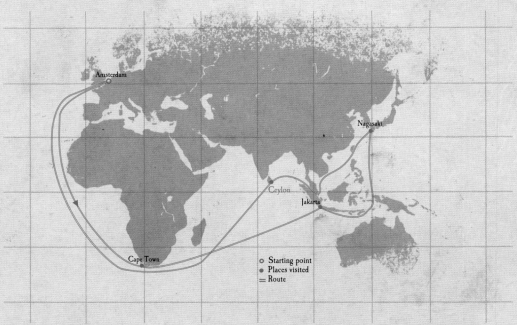

Amsterdam · Nagasaki · Ceylon · Jakarta · Cape Town

○ Starting point
● Places visited
= Route

In the 1770s, a Swede who wanted to go to Japan had to first learn Dutch. The Empire of the Rising Sun was trying to protect itself from the West and only tolerated the arrival of two Dutch ships a year. That is why Carl Peter Thunberg left his native Sweden to spend 3 years in the Dutch colony of South Africa. There, he collected plants with Francis Masson, that "clever English gardener," until he moved on to the next port, Java. Finally, he was able to embark on a Dutch East Indies Company (VOC) ship as the doctor. His secret mission was to "assemble the greatest possible number of seeds, plants, shrubs and trees for the Amsterdam botanical garden."

But in the port of Nagasaki, the Japanese were outraged by the "Europeans' wickedness and their devious ways of smuggling contraband merchandise." Therefore, the cannonballs, powder and cash were placed under seal and trips to shore strictly limited. Thunberg got close to his interpreters who were keen students of medicine, and gradually obtained "permission to go on botanical hunts near Nagasaki, permission that had never before been granted to a European." When this permission was abruptly withdrawn, Thunberg proposed to his interpreters that he would teach them about surgery if they would bring him "everything they could in the way of rare and interesting plants, especially those that grow inland, where foreigners would never be allowed." Later, when the sailors were confined for several months to the tiny island of Deshima, the botanist was reduced to studying the forage given to animals, but even there he found "very rare plants worthy of European herbaria."

His greatest stroke of luck was when he accompanied the Dutch ambassador to see the emperor in Tokyo. For 34 days, Thunberg crossed the provinces of Japan on foot, on horseback, in boats or sedan chairs, sometimes slipping away from his escorts to climb rocks and "gather in a handkerchief rare plants that were just beginning to flower." In the Osaka botanical garden he was able to purchase trees in pots, even maples whose export was forbidden. He was able to smuggle out dozens of species, including some magnificent Japanese lilies — superb, tiger, spotted, heart leaved, Easter — azaleas, sago palm and the famous *Ginko biloba*.

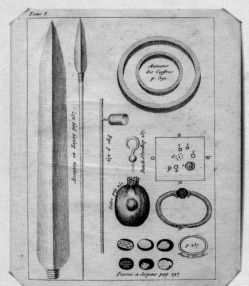

Various objects collected by Thunberg on his journeys. Engraving from Voyages de C.P. Thunberg au Japon par le Cap de Bonne-Espérance, les Isles de la Sonde, etc., *Carl Peter Thundberg, 1796*

TELLING TIME WITH SPICES

Thunberg observed that Japanese night watchmen used little incense boxes in which they burned small, regular amounts of powdered bark of the Japanese star anise (*Illicium anisatum*) as a way to measure the passage of time precisely.

THUNBERGIA: A LARGE FAMILY

The botanical genus *Thunbergia* covers some 200 species that originated in Africa, Madagascar or the Pacific region. They are often climbing plants such as the spectacular black-eyed Susan vine (*Thunbergia alata*) or the blue trumpet vine from China (*Thunbergia grandiflora*).

"February 7 was the first time I took advantage of the new permission the Governor of Nagasaki had given me to gather plants in the outskirts of that city. I was accompanied by a great many interpreters, officers and slaves."

Carl Peter Thunberg

FACING PAGE

HERBARIUM PLATE

Clutia thunbergii (formerly *Clutia daphnoides*)

This specimen was collected on the Cape of Good Hope, South Africa, by Thunberg himself, and mounted in 1779.

Clutytia

Cap de B. Esp. Thunberg. 73.

INSTITUT DE BOTANIQUE
DE MONTPELLIER

Clutytia daphnoides Muell.
Se. XV p. 1056 n° 14

IN THE LAND OF HEATHERS AND SUCCULENTS

Francis Masson, a gardener in South Africa

Francis Masson was born in 1741 in Aberdeen and began his working life as a gardener's apprentice in Scotland, before arriving in London around 1760. There, he was employed at Kew Gardens, where Joseph Banks was the director. In 1772, Banks began sending Masson on plant-hunting expeditions to South Africa, Madeira, the Canary Islands, the Azores and the West Indies; he was the first in a long line of adventurous collectors for Kew Gardens. Masson was captured by the French in Grenada, but freed in 1779. He returned to South Africa in 1786 and stayed there until 1797. He left again for North America in 1797 and collected plants around the Great Lakes. He died, probably of cold, in Montreal in 1805.

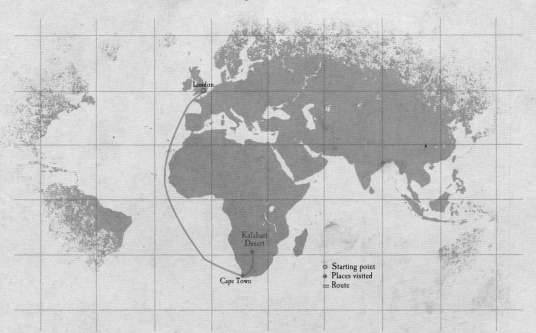

London

Kalahari Desert

Cape Town

○ Starting point
● Places visited
═ Route

"On the 10th of December, 1772, I set out from the Cape Town, towards the evening, attended by a Dutchman, and a Hottentot who drove my waggon, which was drawn by eight oxen; this being the manner of traveling there." And thus did Francis Masson, a gardener from the Royal Botanic Gardens at Kew, set out as the first plant collector for His Majesty, George III, King of England. Joseph Banks, at that time the unofficial director of Kew Gardens and former traveling companion to James Cook, had arranged for the famous captain to take his envoy as far as South Africa. During 3 fruitful years, the enterprising gardener roamed and harvested throughout the Cape region. He went alone on a 4-month expedition and on two others with the Swedish botanist Carl Peter Thunberg. The country was then being settled by the Boers who welcomed them into their farmhouses surrounded by wheat fields, grapevines and orchards. His ox-drawn wagon was slow and heavy, and his road was full of pitfalls. On the banks of the Berg River, Thunberg "plunged over head and ears" into a hippopotamus wallow, and got himself out with great difficulty. The Hottentots and Bushmen were stalking them, and their arrows were poisoned with snake venom or euphorbia juice. The forests were not only full of elephants and buffalo, "a fierce animal, and larger than our English oxen," but also full of unknown trees. That is where Masson found the specimen of Eastern Cape giant cycad that, after a long and perilous voyage, arrived at Kew, where it is still living more than two centuries later.

Discoveries were everywhere; South Africa had an infinite wealth of flora. Northeast of Cape Town, Masson collected "the seeds of many beautiful species of erica (heathers)," iris, gladioli, geraniums, agapanthus, a new variety of amaryllis and *Protea cynaroides* (king protea), with enormous star-

Engraving of Stapelia ambigua *(now called* Gonostemon grandiflorus*) from* Stapeliae novae *or* A collection of several new species of that genus, discovered in the interior parts of Africa, *Francis Masson, 1796*

shaped flowers. Then he left for the burning sands of the Kalahari Desert. In that "thirsty country" the oxen plodded, tongues hanging out, and lions lurked around improbable watering holes. But in an explosion of color, the earth was covered with completely unknown plants: succulents, Ficoidaceae (now Aizoaceae), euphorbia, crassulas, cotyledons, *Aloe dichotoma* with its forking branches, and one of the most flamboyant exotic flowers, the bird of Paradise flower (*Strelitzia reginae*). Crowned with success, Masson returned to South Africa in 1786 during the war between Britain and the Boers. Only able to collect in secrecy, he risked his life for one more South African marvel, seeds of the pure white arum.

FINAL VOYAGE TO CANADA

In 1797, Joseph Banks persuaded Francis Masson to undertake a botanical expedition to North America. The botanist-gardener was first captured by French pirates before being allowed to land at New York. He collected plants along Lake Ontario, and around Toronto and Montreal, went up the Ottawa River and to Lake Superior with agents of the North West Company, and explored the shores of the Saint Lawrence before his sudden death in 1805. He sent Banks seeds and living plants, specimens of wild rice, fruit trees, nut trees and willows.

"We went up to the top Riebeck's Casteel, which is very high, and on the North side inaccessible. It is about four or five miles long, and very narrow on the top; we collected here many remarkable new plants, in particular a hyacinth, with flowers of a pale gold color."

Francis Masson

No 94 *Erica pulchella* Houtt.
Benth in DC. Prod. 7. 662

2 or 3 feet high

sandy places near
Wynberg Cape Div

C. D. Mason
Dec 1839

HERBARIUM
1854.
BENTHAMIANUM

1102 *Erica pulchella* Houtt

Africa australis,
Zeyher,   Wallich, 1847.

HERBARIUM
1854.
BENTHAMIANUM

HERBARIUM
1854.
BENTHAMIANUM

Erica pulchella Houtt
var nuda

Cap.   Hb. Reg. Br.

THE VANISHED BOTANICAL EXPEDITION
La Pérouse and La Martinière in the South Pacific

Jean-François de Galaup, comte de La Pérouse, was born in 1741 in Albi. At 15, he entered the prestigious marine guards school at Brest and began an exceptional career. He held the rank of ensign at 23, lieutenant at 36 and captain at 39. He made a name for himself in the Seven Years War, defending Canada, then accepted a mission to Mauritius, sailing the Atlantic and Indian oceans, before achieving great glory in the American War of Independence. He was a freemason, an excellent sailor and a clever strategist. He was respected by his men as well as by the officers above him because of his great humanity. He died in 1788 in a shipwreck in the Pacific.

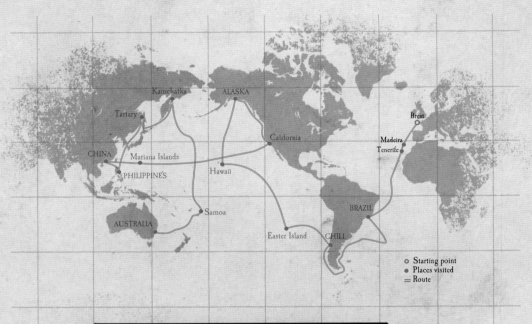

"Monsieur de La Pérouse will use all means to improve the condition of the people he visits, by taking vegetables, fruit and trees from Europe to their lands and teaching them how to sow and cultivate them." Such were the orders from Louis XVI to the count of La Pérouse regarding botany, among many other specific instructions. The king took a personal interest in this prestigious expedition whose goal was to outshine Cook's three voyages. It was expected to last 4 years, cover 93,000 miles (150,000 km) over all the oceans and through all latitudes, assess the fur trade between North America and China, and perhaps even discover the fabled Northwest Passage between the Pacific and Atlantic.

Two solid frigates, *L'Astrolabe* and *La Boussole* carried a crew of elite sailors and the cream of French scientists including a physicist, an astronomer, an engineer, and the young botanist Joseph Boissieu La Martinière. The holds were filled with 3 years of provisions and gifts for indigenous peoples — seeds, potted trees, pearls, muslin, cavalry helmets and medals with the king's likeness. La Pérouse wrote to his wife Éléonore, "Once again, my dearest, do not make any more comments on my expeditions, because you will send me into despair, and all has been decided ... Farewell, my angel, I love you with all my heart and soul." And on August 1, 1785, the expedition set off from Brest.

Louis XVI, King of France, giving his instructions to ship captain Jean-François Galaup, comte de La Pérouse, regarding his voyage of exploration around the world, in the presence of the Marquis de Castries, minister of the navy, June 29, 1785. Painting by Nicolas Monsiau, Museum of the Chateau of Versailles.

By 1786, *L'Astrolabe* and *La Boussole* had visited Easter Island, Hawaii and Alaska. In 1786 and 1787 they reached Macao and the Philippines, sailed between Japan and Korea, and rested a while in the Russian port of Kamchatka. At each stop, La Pérouse noted of the botanist that "M. de la Martinière, with his usual activity, visited the ravines and courses of rivers, to search on the banks for new plants." The expedition left for Samoa and in January 1788 made landfall at Botany Bay in Australia. While the naturalists were collecting plants, gathering seeds of *Banksia*, for example, the captain entrusted his logbook to an English frigate. That was the last time he was heard from, and the beginning of a mystery. In January 1783, as Louis XVI awaited execution, he asked, "Have we heard anything from La Pérouse?"

Read the next parts of the La Pérouse mystery on pages 80 and 100.

AN ADVENTURE WITHIN THE ADVENTURE

The expedition's Russian translator, Barthlémy de Lesseps, disembarked in Kamchatka, entrusted with La Pérouse's first log book. Only 22, he traveled for more than a year, crossing Siberia by sled, then Russia and Germany, to deliver it to Versailles.

WAS LA PÉROUSE AN ANTI-IMPERIALIST?

As a humanist of the Enlightenment, Lapérouse commented, "I did not think it my duty to take possession of it in the name of the king: the customs of Europeans are in this respect completely ridiculous. ... regarding as an object of conquest a land, which its inhabitants have watered with their sweat, and which during so many ages has served as a tomb to their ancestors."

"The Indians demanded one glass bead for every plant Mr. de la Martinière collected and they threatened to beat him if he refused to pay this compensation; when the massacre happened, he swam to our canoes in a hail of stones, his sack of plants on his back, and thus was able to save them."

Jean-François Galaup, comte de La Pérouse

FACING PAGE

HERBARIUM PLATE

Banksia ericifolia L.f.
This specimen of heath-leaved banksia was collected in Australia.

B. ericifolia. L.f.

FLORA AUSTRALIENSIS.
named by Mr. BENTHAM.

V. 547.

Banksia
  ericifolia, Sieb.

Hastings R.
  N. S. W.
Hb. Oldfield.

IN THE SHADE OF AMERICAN TREES

André Michaux, a French nurseryman in the United States

André Michaux was born in 1746 on a farm on the royal estate of Versailles, where his father was the manager. He was passionate about agronomy and went to Paris to study botany. He traveled through France and England, and became friends with Jean-Baptiste Lamarck, a naturalist, and André Thouin, head gardener in the Jardin du roi. Employed as the secretary to the French consul in Persia, he crisscrossed that country for 2 years and created such a fine herbarium that Louis XVI named him the "botanist of the nurseries" and sent him to the United States in 1785. His son François-André continued his father's work observing and collecting species of the Americas.

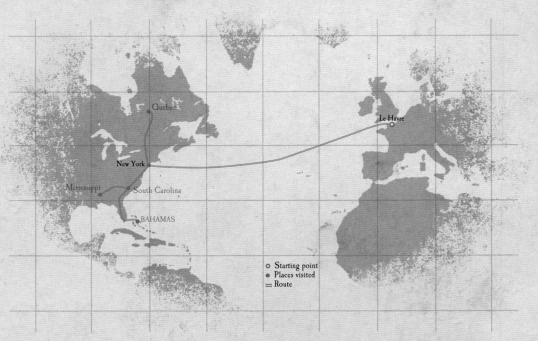

Photograph of his son, François-André Michaux.

In 1785, France's forests were depleted from supplying the shipyards during decades of war with England. Louis XVI was officially worried. Luckily, France had an excellent relationship with the young United States of America, where the forests were still untouched. Benjamin Franklin and Thomas Jefferson, who had been impressed with the Jardin du roi, enthusiastically welcomed the idea of taking an inventory of this natural resource, and André Michaux was the man for the job. He was 39 and had just come back from travels in Persia where, although he risked his life many times, he had fallen in love with the exotic vegetation. "I cannot tell you how happy I was to visit the countryside. Considering the multitude of plants covering the prairie, I was often dazzled and had to force myself to calm down for a few moments. I could not sleep at night; I was impatient for the next morning."

Michaux quickly left for New York with his son, François-André, who was 15. Right away, he created a nursery in New Jersey and another in South Carolina. He went up and down the east coast, to the Bahamas and northern Quebec, into the heart of the continent as far as the Appalachians and the Mississippi. Guided by the native people, he explored rivers in a birch bark canoe,

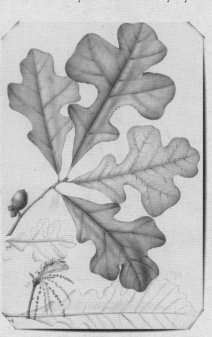

Engraving of Quercus obtusiloba *from* Histoire des chênes d'Amérique... *by André Michaux, illustrated by Pierre Joseph Redouté, 1801*

lived like a trapper in the wildest of lands, despite both heat and cold, and never forgot to write in his journal. "We almost always camped in the forest, and my father wrote by firelight, noting the observations he had made during the day," wrote his son.

In 11 years, Michaux sent 1,700 North American plant species to France in the form of seeds (90 crates) or seedlings (60,000 in pots): basswood, ash, yucca, maple, oak, pine, cypress, walnut and more. It was Michaux that gave Europe the woodbine or Carolina jasmine (*Gelsemium sempervirens*), spectacular magnolias and *Rhododendron catawbiense*, from which many hybrids have been formed. But the French Revolution was a disaster for Michaux, who was a dedicated republican. The new government refused to pay him and he had to return to France. His ship sank; his crates of seeds were lost and his herbarium was soaked. But there was worse to come: the beautiful nurseries at Rambouillet, where thousands of his plants were flourishing, were destroyed. Stubbornly, Michaux proposed returning to the United States but had to sign on as the naturalist for Baudin's expedition to Australia. On the way, he stopped at Madagascar and died of fever there in 1802, without ever seeing the green forests of America again.

A SENSATIONAL MAGNOLIA

In Tennessee, Michaux discovered a magnolia with large, shiny leaves and enormous, fragrant, white flowers with long red pistils. Empress Josephine was the first person in Europe to possess this plant, which became a sensation under the name *Magnolia macrophylla* or *Magnolia tripetala*.

AN UNINTENTIONAL SPY

In 1793, Citizen Genet, the new French ambassador to the United States, was pushing the Americans to go to war with Spain. Under cover of his botanical expedition, Michaux was to carry secret messages to Kentucky, but the plot was uncovered and the botanist was greatly relieved.

"On the mountain top we found very solid white ice and the streams were full of icicles ... We went down a number of rapids without portaging and after traveling 15 leagues we arrived at the Grand Rapids. ... We camped near the first white pines (Pinus strobus or Weymouth pines) that one encounters when going down from Mistassin."

André Michaux

FACING PAGE

HERBARIUM PLATE

Magnolia cordata

This specimen was collected by Michaux himself in 1787.

Magnol. cordata

Georgie — Michaux
1787

THE ADVENTURES OF THE BOUNTY AND THE BREADFRUIT TREE

Captain Bligh and the Tahiti mutineers

Born in 1754 to a wealthy English family, William Bligh was signed on to the Royal Navy at 9 years old. He was a gifted cartographer and rose quickly through the ranks to become captain of the *Resolution* at 23. On that ship, he sailed with James Cook on Cook's third and final voyage. After the mutiny of the *Bounty*, Bligh was exonerated and continued his career but encountered — or provoked — many disciplinary problems. He was victorious in two naval battles against Napoleon, then was governor of the colony of New South Wales, Australia, where he put down a bloody rebellion. Bligh finished his career as vice-admiral and died in 1817.

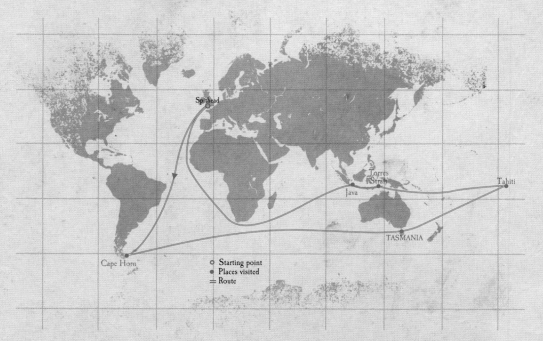

○ Starting point
● Places visited
═ Route

In 1772, during James Cook's first visit to Tahiti, botanist and explorer Joseph Banks noted the breadfruit tree, whose starchy fruit was very nourishing. In 1787, when he had become director of the Royal Gardens at Kew, and the British were cut off from cheap wheat supplies, while their slaves in the Caribbean were starving, Banks remembered the breadfruit. He began the first attempt at acclimatizing plants on a grand scale, from one end of the earth to the other. A vessel was specially equipped, with an immense space between decks reserved for the potted plants. There were hatches for air flow and glassed windows for sunlight, a floor with holes and a system for recycling water. The ship was called the *Bounty* and its command was given to the finest navigator in the Royal Navy, Captain William Bligh. A cartographic genius, he had proved his talent in Polynesia alongside Cook. His second in command was Fletcher Christian, an officer he trusted completely. His friend David Nelson, a botanist from Kew, was chief scientist.

Unfortunately, all these preparations took a long time and the *Bounty* missed the best season for rounding Cape Horn. After a month spent fighting blizzards, Bligh gave up on the South American route and set his sails for Africa. Crossing the Atlantic and the Indian oceans, he finally reached Tasmania; it had been a 10-month voyage and the crew's nerves were frayed. Haughty and demanding, Bligh ruled his men through deprivation and whipping, even more than the cruel code of naval justice then allowed.

The *Bounty* reached Tahiti in October 1788. Bligh gave presents to the Tahitian king Tinah and invited him to offer King George some breadfruit trees. David Nelson supervised the Tahitians potting up these trees. Bligh congratulated himself on collecting all eight species of breadfruit on the island, along with several other Polynesian fruits such as, "the avee, which is one of the finest flavoured fruits in the world; the ayyah, which is not so rich, but of a fine flavour and very refreshing; the rattah, not much unlike a chestnut ..." Nelson also established a garden to acclimatize the seeds gathered at the Cape of Good Hope. Bligh distributed fruit-stones and almonds for planting among the chiefs, and "as they are very fond of sweet-smelling flowers, with which the women delight to ornament themselves, I gave them some rose-seed." It took six months to do this work, and the crew felt they were in paradise.

(continued on page 78)

Mutiny on the Bounty: *The mutineers turning Lt Bligh and some of the officers and crew adrift from His Majesty's Ship* Bounty, *April 29, 1789. Hand-colored aquatint by Robert Dodd*

CAPTAIN COOK'S LAST VOYAGE

William Bligh and the botanist David Nelson had both been part of Cook's third and last voyage, when Cook was killed by natives on the Big Island of Hawaii in 1779. Lieutenant Bligh was perfecting his navigational skills on the routes of the Pacific as master of the *Resolution* and David Nelson was a junior officer. When they were in Tahiti in 1777, Nelson planted grapefruit trees and when he returned on the *Bounty*, 20 years later, he found two of these trees, "covered with unripe fruit."

"Wednesday, the 5th. The weather variable, with lightning, and frequent showers of rain. Wind E.N.E.

This was the first day of our beginning to take up plants; we had much pleasure in collecting them, for the natives offered their assistance, and perfectly understood the method of taking them up and pruning them."

William Bligh

FACING PAGE

HERBARIUM PLATE

Artocarpus altilis

This specimen of breadfruit was collected by Forster in the Pacific islands.

Artocarpus incisa

Tahiti
Forster

= Artocarpus Mli (Park.) Fosb.

ISO-

Nᵒ 3661

Vu pour la flore dè Polynésie Française

DATE : 4.1970

DET : J. Florence

Society Islands

Artocarpus altilis (Park.) Fosberg

DET D. Ragone   19 90

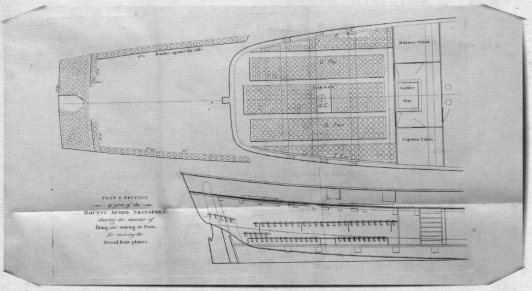

Plan and profile of the deck of the Bounty, *fitted for receiving the breadfruit plants. Drawing from* A voyage to the South Sea undertaken by command of His Majesty, for the purpose of conveying the bread-fruit tree to the West Indies, in His Majesty's ship the Bounty, commanded by Lieutenant William Bligh, *William Bligh, 1792*

The men enjoyed a soft life on these leeward islands, so much so that three of them deserted. They were captured and whipped.

On March 31, 1789, the hold full of 1,015 flourishing breadfruit trees, the *Bounty* was ready to set out on the sea. The men were reluctant to leave, especially Fletcher Christian who had been charmed by Tahiti. Bligh tried to reinstate his iron discipline. The atmosphere was stormy and getting more combative. On the morning of April 28, Christian and some sailors took hold of Captain Bligh, and threw him and 18 others into a ship's boat with some sails and ropes, 150 pounds of bread, a barrel of water, a little wine and rum, four sabers and a compass, but no map or sextant.

Bligh accomplished an incredible feat of navigation: 5,178 miles (8,334 km) across the Great Barrier Reef and the Strait of Torres to the Dutch colony of Timor. In a vessel 23 feet (7 m) long, overloaded and without a deck, the men had to take turns sitting and lying down. They survived on a mouthful of water and a piece of bread equivalent to the weight of a pistol shot, weighed out on Bligh's homemade scale made of coconut shells. One of them died when they were landing on an island, but many others, including David Nelson, died of fever after they reached Timor. Bligh hurried home to England where he faced a court martial for losing his ship. He was quickly exonerated, especially with the help of Joseph Banks who sent him on a second expedition. In 1792 he took command of the *Providence* and successfully transported 3,100 breadfruit plants to Jamaica.

The *Bounty* mutineers met a variety of fates. They all agreed to throw Bligh's precious plants overboard, but disagreed on the strategy to follow next. Having failed to colonize one island, some returned to spend a few happy months on Tahiti. But they were found by the *Pandora*, which took them to London, despite a shipwreck and other misadventures. Three of them were hanged and three were pardoned. Others took refuge on remote Pitcairn Island, where they eventually killed each other off, although their fair-skinned, blue-eyed descendants were found there 20 years later.

Bligh's breadfruit trees did not thrive in Jamaica but survived in the French Caribbean. The African slaves found the fruit distasteful and refused to eat it.

Drawing of a breadfruit from A voyage to the South Sea undertaken by command of His Majesty, for the purpose of conveying the bread-fruit tree to the West Indies, in His Majesty's ship the Bounty, commanded by Lieutenant William Bligh, *William Bligh, 1792*

"The boatswain came to my hammock and waked me, telling me to my great surprise, that the ship was taken by Mr. Christian. I hurried on deck and found it true — seeing Mr. Bligh in his shirt with his hands tied behind him and Mr. Christian standing by him with a drawn bayonet in his hand and his eyes flaming with revenge."

James Morrison, Master Gunner

IN THE LAND OF EUCALYPTUS
La Billardière searches for La Pérouse

1760

Jacques Julien Houtou de La Billardière was born in 1755 to a wealthy family in Alençon. He was one of the Enlightenment scientists who believed in "enlightened reason" and the idea of liberty. After his medical studies, he studied botany "in the great book of nature" among Joseph Banks's exotic collections in England, then in the Alps and Asia Minor. In 1791 the Constituent Assembly proposed that he follow the trail of La Pérouse. La Billardière, then 36, "eagerly seized this opportunity to explore the South Seas" and devoted the latter part of his life to writing the story of that voyage and introducing eucalyptus trees to Europe. He died in 1834.

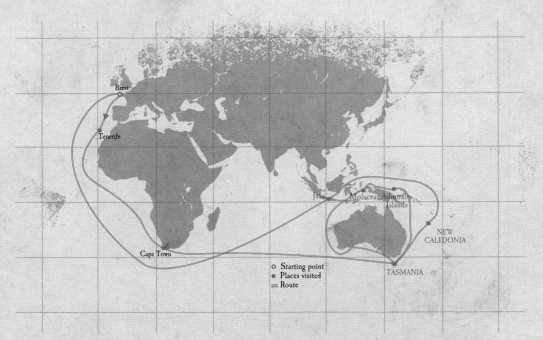

Brest
Tenerife
Cape Town
Java Moluccas Admiralty Islands
NEW CALEDONIA
TASMANIA

○ Starting point
● Places visited
═ Route

Three years had passed without any news from the Pacific about La Pérouse, so it was decided to send a search party under the command of Admiral d'Enrecasteaux. Along the way, he was expected to "enrich France's flora." In September 1791 *La Recherche* and *L'Espérance* set out, laden with hope, pots for the plants and gifts for the natives.

The expedition began well. Dropping anchor at Cape Town, d'Entrecasteaux learned that "natives in French naval uniforms" had been seen in New Guinea. But 2 months later his ships suffered great damage as they arrived in Tasmania. The botanists took advantage of this forced delay: Lahaye, a gardener, created a garden with his European seeds; Jacques Julien Houtou de La Billardière went gathering in the forests, met "savages" and made thousands of discoveries. He was thrilled by the eucalyptus and discovered the blue gum tree, *Eucalyptus globulus*. In Tonga he harvested 200 breadfruit trees but he was frightened by the cannibals of New Caledonia.

For the better part of 1793, the two ships sailed among the numerous Admiralty Islands. Scurvy and the incessant tropical rain were ravaging the crew. One of the many tiny islands they passed was one that d'Entrecasteaux named "La Recherche" without knowing that *La Boussole* and *L'Astrolabe* lay at the bottom of its lagoon. Their already difficult task was complicated by politics — class warfare between aristocratic officers and the commoners led by La Billardière, a fervent supporter of revolutionary ideas. Admiral d'Entrecasteaux died of scurvy and the expedition eventually reached Java. Some surprises awaited them there: Louis XVI had been guillotined and Holland had declared war on France! The royalists on the ship gave the republicans to the Dutch, and La Billardière was thrown in prison.

Engraving of Eucalyptus globulus *from* Voyage in search of La Pérouse: performed by order of the Constituent Assembly, during the years 1791, 1792, 1793, and 1794, *Jacques Julien Houtou de La Billardière, 1799 (English translation, 1800)*

His precious collections were confiscated, and he just had time to send 11 breadfruit trees with Lahaye who planted them on Mauritius and then at Cayenne, where they flourished. But La Billardière's collections were still traveling: the Dutch vessel that was carrying them was captured by an English frigate and they ended up in the hands of Joseph Banks at Kew. Nobly, he refrained from opening the crates and sent them to his unfortunate colleague who by then was back in Paris, refusing to take "a single botanical idea from a man who searched them out while risking his life."

HARD LIFE ON BOARD SHIP

The job of naturalist on these grand expeditions was no easy job. La Billardière complained that his herbarium paper was always damp and that he had to dry it in the oven. His tiny cabin was already full of plants, minerals and dead birds, so he had to store his specimens in the "great room," but a lieutenant thought them a nuisance and threw them out. As for food, the naturalists tried in vain to take their fair share of provisions on their excursions but only got ship's biscuit, some cheese and salt pork.

"That cannibal told us that the flesh of arms and legs was cut in slices seven or eight centimetres thick, and that the most muscular parts were a very agreeable food for these people. That made it clear to us why they often pressed our arms and legs with a violent and hungry attitude."

Jacques Julien Houtou de La Billardière

Read the final installment of the mystery of La Pérouse on page 100

FACING PAGE

HERBARIUM PLATE

Eucalyptus globulus

This specimen of the blue gum tree was collected in Sydney, Australia, by La Billardière himself, and the specimen plate created in 1839.

SHEET 2 ISOTYPE
of *Eucalyptus globulus*
              Labill.
     J. Chippendale 16/3/1973

Botanical Name *Sucker foliage ♂*
Euc: globulus, Labill:

Examined—J. H. MAIDEN,
              Botanic Gardens, Sydney.

Date *9/61*

81140—2

HERBARIUM
1854.
BENTHAMIANUM

1760

FOLLOWING BONAPARTE TO THE LAND OF LOTUS AND PAPYRUS

Alire Raffenau-Delile and the Egyptian campaign

Alire Raffenau-Delile (known as Delile) was born into a family of courtiers at Versailles in 1778 and first studied botany in the Trianon gardens. Following the revolutionary turmoil. he studied medicine in Paris and, at 20, was recruited for Napoleon Bonaparte's expedition to Egypt. After that, Delile left for the United States where he was a merchant and then a practising doctor. He was recalled to France in 1807 to work on the flora included in the monumental *Description de l'Egypte*. Delile obtained the position of head of botany at the University of Montpellier in 1810, after the Empire fell. He died there in 1850, leaving a rich legacy of some 60 books on medicine and botany.

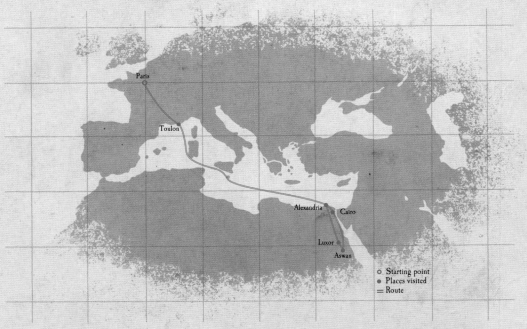

o Starting point
● Places visited
= Route

On May 19, 1798 (known in the Republican calendar as 30 Floréal, year VI), an armada left Toulon; 38,000 soldiers and 160 scientists departed for an unknown destination. Napoleon wanted to strike a surprise blow against England and did not reveal his target until they were at sea. The target was Egypt, the hub of Britain's commerce in the Orient. The young general had decided to add an exceptionally wide-ranging scientific mission to the military campaign. He asked his friend Gaspard Monge, a mathematician, to recruit the intellectual elite of the day: engineers, architects, astronomers, geographers, historians, surgeons, painters, printers and, of course, naturalists. Among these brilliant minds was Alire Raffenau-Delile, botanist and physician, age 20. He traveled with the campaign as far as Aswan in Upper Egypt, then to Mount Sinai in Palestine.

It was a difficult beginning. The army had to face a scorching and desolate plain and although the scientists had lost most of their instruments, they set to work with great inventiveness and results. Delile learned Arabic, decoded inscriptions on obelisks using his knowledge of Greek, "inspected the tombs to find what grain had nourished the pharaohs," collected plants all along the Nile Valley as the French troops advanced, harassed by the Mamluks (Mamelukes). Their guerilla tactics led to an amusing order from a veteran of the Italian campaign: "Here come the Mamluks. Form a square. Put the donkeys and the scientists in the middle!"

Delile was soon named director of the Cairo botanical garden, and he enriched it with his collections: lichens, many desert succulents, pink lotus, papyrus and others. He also wrote articles on the doum palm and the endemic flora of the Middle East for the scientific journal, *La Décade égyptienne*, which was printed in Cairo on presses brought from France, and formed the bases for the 20-volume *Description de l'Egypte* published during the Second Empire.

Although the scientific expedition was an incomparable success, the military campaign was not. The English wiped out the French fleet at Abukir and Bonaparte failed in his siege of Acre. He quickly went home to France, leaving the army, which capitulated in 1801. In Alexandria, the scientists were forced by the English to hand over their collections. They said they would prefer to burn them themselves, and thus were able to conserve the most important parts. Delile was able to return to France with a large herbarium and precious lotus and papyrus plants that he successfully acclimatized in the Montpellier botanical garden.

THE KEY TO HEIROGLYPHICS

In the summer of 1799, digging near Rashid (Rosetta) revealed a basalt stone 3 feet (1 m) tall. Three inscriptions were engraved on it, in hieroglyphs, in Demotic and in Greek. It soon became apparent that the text of all three was the same. The stone was sent to the Institute of Egypt in Cairo, where Delile had a mold and three copies made. One of them was quickly sent to France, but the original Rosetta Stone was captured by the British in 1801. Delile's copy made it possible for Champollion to solve the mystery of hieroglyphics in 1822.

The scientists of the Egyptian commission, preparatory sketch for the Déscription de l'Egypte

"There are lichens in the highest part of the desert, between Cairo and the Red Sea. They cover the dry stones and only the greatest amount of decay can destroy them; when there is fog, they are reborn. These same lichens are found near the summit of the pyramids of Giza, only on the north side, and on the ones at Saqqara."

Alire Raffeneau-Delile

FACING PAGE

HERBARIUM PLATE

Nigella

Raffenau-Delile brought the seeds of these specimens of *Nigella* back from his expedition to Egypt and planted them at Versailles.

nigella

habba soda (Arab)
de graines d'Egypte semées
à Versailles à mon retour.

FROM MANGROVES TO YERBA MATE

Humboldt and Bonpland in Central America

Baron Friedrich Wilhelm Heinrich Alexander von Humboldt was born in Berlin in 1769, the son of a rich Prussian officer. He had an excellent education in history, natural science and philosophy, and became a mining engineer. At 18, he inherited a huge fortune and with his brilliant mind, decided to devote himself to studying the "unity of nature." Humboldt became friends with Georg Forster, the botanist from Cook's second voyage, and together they traveled across Europe. Later, he spent 5 years in Central America with Bonpland. In 1829 he crossed the Russian empire, but his greatest ambition was to publish the accounts of his many voyages, adding up to 30 volumes. He died in 1859, at the age of 90.

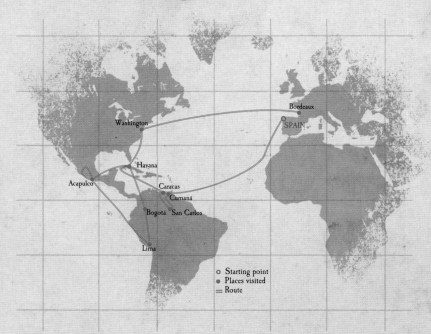

○ Starting point
● Places visited
═ Route

Aimé Jacques Alexandre Goujaud was born at La Rochelle in 1773. According to the family legend, his grandfather who worked in a vineyard said of him, "That boy will be a good plant (bon plant)," and so he was known as Bonpland. He studied at the college in Rochefort to be a naval surgeon, but was more interested in botany, which he studied with the celebrated André Thouin at the Jardin des plantes (formerly the Jardin du roi) in Paris. There, he was fascinated by the acclimatization of exotic plants. Because of his talents, he was chosen to take part in Bougainville's planned second expedition, but it was cancelled and Bonpland ended up leaving with Humboldt. When he returned, he became the Intendant of Empress Josephine's gardens at Malmaison, then left once more for Argentina, where he died in 1858.

"A dazzling light was spread along the white hills clothed with cylindrical cactuses, and over the smooth sea, the shores of which were peopled by pelicans, egrets, and flamingoes. Every thing announced the magnificence of nature in the equinoctial regions." In July 1799, at the mouth of the Manzanares River, Alexander von Humboldt and Aimé Bonpland were dazzled by Venezuela even before they landed.

The two friends had been preparing a long time, thanks to Humboldt's fortune, and had the latest in scientific instruments. But it required all the charm of the aristocratic, polyglot von Humboldt to convince King Charles V of Spain to authorize the voyage and finally be permitted to cross the Atlantic on a ship ravaged by typhoid fever.

"The first plant we gathered on the American continent was the *Avicennia tomentosa*, which barely reaches a height of two feet in this place." That white mangrove was the beginning of innumerable discoveries. In the early days, the two explorers were overwhelmed and "wandered about like madmen." Bonpland said he would "lose his head if these wonders do not soon cease."

After a stay in Caracas, they set out to resolve a century-old scientific dispute, whether the Orinoco and Amazon rivers were independent of each other or whether they were connected by a body of water. Humboldt and Bonpland crossed immense hot, desolate landscapes before entering the forest. As they followed the ever-diminishing water course, they had to abandon their large dugout canoe for smaller vessels. Their 1,550 mile (2,500 km) journey included being carried on men's backs, terrifying rapids, near-drownings from which the men, books, herbarium specimens and instruments came out soaked,

Drawing of the Michoacan Indians in native costume, from Vues des cordillères et monuments des peuples indigenes de l'Amérique, *by Humboldt and Bonpland, 1810*

GEOGRAPHY OF PLANTS

In Colombia, Bonpland and Humboldt climbed Chimborazo volcano, doing a painstaking survey of the vegetation in relation to altitude, climate and topography. This became the basis of a new scientific discipline, phytogeography.

A COLD CURRENT

In 1802 Humboldt and Bonpland sailed from Lima to Acapulco. Never idle, Humboldt took regular temperature readings of the ocean. Thus he discovered the cold current that comes from Antarctica and goes up the west coast of Latin America; it now bears his name.

nights being threatened by jaguars, snakes, crocodiles and curare-poisoned arrows, and, of course, the cruel torment of mosquitoes. "Mr. Bonpland and myself tried the expedient of rubbing our hands and arms with the fat of the crocodile and oil of turtle-eggs, but we never felt the least relief." But they carried on, making maps, measuring altitudes and temperatures, observing the stars, collecting plants, insects, animal carcasses and rocks. Finally, in May 1800, they reached the Casiquiare canal that connects the watersheds of the two great rivers, via the Rio Negro.

(continued on page 88)

"For three months I slept outside, in the forest surrounded by tigers and terrible snakes, or on beaches covered with crocodiles. Bananas, rice and manioc were our only food, since all our provisions rotted in that damp and hot country. Nature in these mountains is majestic!"

Alexander von Humboldt

FACING PAGE

HERBARIUM PLATE

Verbena strigosa Cham.

These specimens of verbena were collected at Sao Paulo, Brazil, by Humboldt himself and mounted in 1836.

Verbena strigosa Chamisso
i Linnaea 7. 256.

Brasilia meridionalis
15 NOV 1907

Ex reliquiis Sellowianis. Humboldt
3869   ded. 1836

Verbena strigosa Cham

Brasilia meridionalis.
S Paulo   15 NOV 1907

Ex reliquiis Sellowianis. Humboldt
1739   ded. 1836

It was an immense success: the century-old mystery had been solved and they harvested 20,000 botanical specimens including *Brugmansia* or angel's trumpet, orchids, passion flowers, zinnias and fuchsias. But they were exhausted and Bonpland suffered from fevers for a month. They sent their collections to Paris, in several shipments so that at least one part would arrive intact. That was a wise precaution because one of the shipments sank and another was captured by the British.

In 1801 they traveled to Cuba and then to Colombia, intending to scale the Andes. They crossed the cordillera and ascended several volcanoes. On the flanks of Chimborazo, they definitively broke the record for altitude — 18,300 feet (5,878 m) — and became gravely ill from lack of oxygen. Nevertheless, they completed an extremely precise survey of the vegetation. Late in 1802, they took a ship to Acapulco, crossed Mexico, and worked hard every step of the way. In 1804, they were received by President Jefferson in Washington and then returned to a hero's welcome in France.

There, they went their separate ways. Bonpland returned to his family at La Rochelle and Humboldt went straight to Paris and then Berlin, gave lectures, and attended salons. They each began to publish their own discoveries. Humboldt took on geography, astronomy and zoology, whereas Bonpland concentrated on botany, based on the 6,000 specimens he had collected, which included 3,800 that were yet unknown to science. Empress Josephine put him in charge of her prestigious gardens at Mailmaison, but

Raft on the Guayaquil River, drawn by Marchais, based on Humboldt's sketches. Taken from Atlas pittoresque: vues des Cordillères et monuments des peuples indigènes de l'Amérique de Humboldt et Bonpland, *1810*

he lost his job when the Empire fell. Bonpland and his young wife Adeline went to live in Buenos Aires, where he continued his botanical activities and, among other things, discovered the secret of *yerba mate*, the Indians' tonic. The seeds of that plant are so tough that they have to be broken down before they will germinate. Bonpland argued for the commercial cultivation of yerba mate, bought land near Paraguay and got tangled up in diplomatic intrigue over the placement of the border. The Paraguayan dictator held him prisoner for 9 years, during which Bonpland farmed, raised cattle, remarried and had two children. The dictator eventually sent him away but separated him from his family. Despite these hardships, Bonpland never stopped collecting plants and sending them to the Muséum in Paris, among them cactus, water lilies and vines. Humboldt, who sometimes spoke unpleasantly about his "dear, best friend" wrote to him, "We should never have parted."

GÉOGRAPHIE DES PLANTES ÉQUINOXIALES.

Tableau physique des Andes et Pays voisins

"Tableau physique" from Géographie des plantes équinoxiales. Tableau physique des Andes et Pays voisins. Dressé d'après des observations et des mesures prises sur les lieux depuis le 10° degré de latitude boréale jusqu'au 10° de latitude australe en 1799, 1800, 1801, 1802 et 1803, by Humboldt

"Humboldt and I lived together like friends or brothers; what was his was mine and what was mine was his. The harmony we had preserved during our long time together helped us forget the multitude of troubles we had suffered among the savages of the Orinoco, Negro and Amazon rivers and on the icy peaks of the Andean cordillera."

Aimé Bonpland

Mus. bot. Berolin.

Recd. May 1893

Leandra pulchra Cogn.

Musco Botanico
Berolinensi

Inter Victoria et Bahia.

Hb. Kunth.

Ex reliquiis Sellowianis. Humboldt
ded. 1876

1770

FLORA OF THE SOUTHERN HEMISPHERE: THE NATURALISTS' TRUCE

Baudin vs. Flinders in Australia

Nicolas Baudin was born in 1754 in Saint-Martin-de-Ré and entered the navy at 21. He took part in the American War of Independence, then engaged in trade with Mauritius and became a supplier of plants to the Emperor of Austria. In 1796 Baudin made a successful voyage as a naturalist to the Caribbean, then set off on an expedition to Australia. He died on the return journey in 1803.

Matthew Flinders was born in Lincolnshire in 1774 and served as a seaman on Bligh's second voyage to Tahiti in search of breadfruit trees. In 1801 Flinders went to map Australia—and to supervise Baudin. He returned to England in 1809 and died at 40, just before his narrative of his voyage was published.

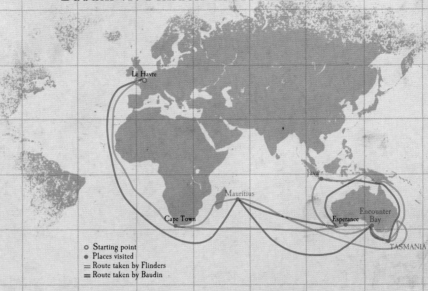

Le Havre

Java

Mauritius

Cape Town   Esperance   Encounter Bay

TASMANIA

○ Starting point
● Places visited
═ Route taken by Flinders
═ Route taken by Baudin

In 1800, the Kingdom of Great Britain and Napoleon Bonaparte's France were still at war. Despite the ongoing conflict, Joseph Banks, President of the Royal Society, was able to arrange a letter of passage for two French ships to Australia, to do scientific exploration for the Institut de Paris. Thus, *Le Géographe* and *Le Naturaliste* left Le Havre in October 1800, commanded by Captain Nicolas Baudin. But the English had already claimed Australia and wanted to keep watch on the French, so in July 1801, Baudin was followed by Captain Matthew Flinders at the helm of the *Investigator*.

The French expedition got off to a poor start. Captain Baudin was exasperated with the demands of the scientists and the dilettantes among the junior officers. When they stopped at Mauritius, 40 scientists and sailors declared they were sick and left the ship. The crossing to Australia was another trial, as it became obvious that *Le Naturaliste* was slow. Food ran short and scurvy appeared. The expedition reached land in June, but a storm separated the two ships. *Le Géographe* went up the Australian coast to Timor to get provisions and take care of the many who were ill. But it was a deadly journey, and young Antoine Guichenot was the only gardener left alive. Although he was "poorly educated," he worked hard and accomplished a great deal for botany.

After 3 months of separation, the two ships set out for Tasmania and, on April 8, 1802, encountered the English ship in a harbor. Flinders and Baudin were cordial, although they did not know that a peace treaty had been signed. Then the French ships continued to Port Jackson (which later became Sydney), where Baudin wrote to Jussieu, "As I write, all the countryside is in bloom and nothing can rival its beauty." He sent *Le Naturaliste* back to Paris with the richest collec-

tions ever received at the Muséum d'histoire naturelle: 24 crates of herbarium specimens, 70 pots of plants, planks of eucalyptus and casuarina, and live animals: kangaroos, black swans, dingoes, emus, snake-necked turtles and parrots. Baudin received no recognition for this exploit. He died in 1803 and his enemies were all eager to tarnish his reputation.

The English expedition went all around Australia. Naturalist Robert Brown brought back a fabulous harvest of 4,000 plant species, of which 1,700 were unknown. But the conflict between Britain and France had heated up again and Captain Flinders was held prisoner for 6 years on Mauritius, which was enough time for the French to start using some of his discoveries.

First page of Nicolas Baudin's logbook, 1799

SPROUTING THE SEEDS

The seeds and plants collected by the Baudin expedition were shared by three gardens: the Muséum in Paris, the Montpellier botanical garden and Empress Josephine's gardens at Malmaison, where they were in the expert hands of Aimé Bonpland, who successfully grew New Zealand flax, eucalyptus and mimosa.

BROWNIAN MOTION

Robert Brown, the Flinders expedition's botanist, used a microscope and was the first to describe the nucleus of cells. In 1827, he examined primula pollen and discovered that, in suspension in water, these tiny particles moved in every direction; in physics that movement is still known as Brownian motion.

"Our head gardener Riedlé, to whom every instant is precious, has found 70 varieties of plants on this little island, and most of them are probably unknown to botanists. Maugé, the zoologist ... has collected 10 species of bird, which he thinks are new. The others have captured kangaroos and some lizards."

Captain Baudin

FACING PAGE

HERBARIUM PLATE

Philotheca calido

This specimen was collected at Port Jackson, Australia, by Baudin himself.

2

Philotheca Gaudichaudii Don

Rec.? 12/80.

Cap. Band.

NOUVELLE-HOLLANDE. *Port Jackson*

Herb. Mus. Paris.

apparently identical with P. australis, Rudge

(R. A. Roly)

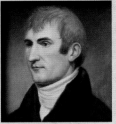

AT THE ROOT OF WESTERN EXPLORATION

Meriwether Lewis crosses the United States

Meriwether Lewis was born to a family of planters near Charlottesville, Virginia, in 1774. He learned at an early age to ride, hunt and survive in the wilderness. At 20, he joined the army during the wars against the Indians. He served under William Clark and became a captain at 26. Both of them respected the Indians and learned their languages and customs. In 1801 Lewis became an aide to President Thomas Jefferson, and was his confidant in preparing the transcontinental expedition in secret. When he returned, he was made governor of the Louisiana territory, but he died in 1809, either murdered or by his own hand.

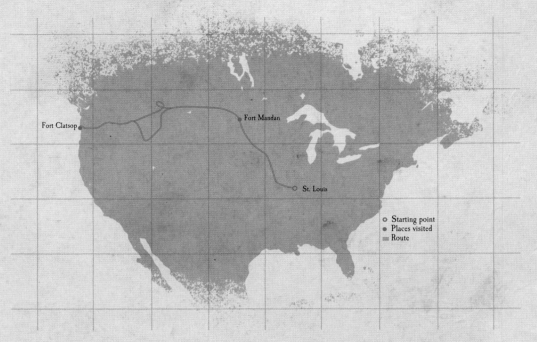

Fort Clatsop   Fort Mandan

St. Louis

○ Starting point
● Places visited
≡ Route

On July 4, 1803, U.S. President Thomas Jefferson revealed that the United States had purchased the province of Louisiana from France. This immense territory stretched from the Gulf of Mexico to Canada, and now it needed to be explored, all the way to the Pacific. Jefferson took personal responsibility for this first transcontinental journey and selected the leaders: his own secretary, Meriwether Lewis, and William Clark. They were good friends and both were army veterans. They took accelerated training to learn how to draw maps of these unexplored regions and evaluate the natural resources. Lewis also took courses in botany, mathematics, paleontology and medicine.

In May 1804, they left St. Louis and went up the Mississippi River. Some 40 men embarked on a shallow-draft keelboat and two canoes. Clark was in command of the boats, while Lewis rode along the riverbank on a horse, taking notes and collecting plants. He collected hundreds of new plants this way, including bearberry, red osier dogwood (which was smoked like tobacco), snowberry, holly-leaved barberry or Oregon grape, bulbs of the blue quamash, wild roses and the Douglas fir. He buried two lots of his specimens on the banks of the Missouri, intending to retrieve them on the return journey, but they were washed away by floods.

The expedition spent the winter of 1804 at Fort Mandan, North Dakota, on the banks of the frozen Missouri. In spring, they set out for parts unknown, meeting Sacagawea, a Shoshone Indian woman who would be of great help to them. In July 1805, the whisky was running short. In August, the explorers took possession of an Indian's bag and found three edible roots in it: valerian, bitterroot and Western spring beauty (*Claytonia lanceolata*). In October, the starving men ate their horses. But they still had hundreds of rapids to run or portage around, including the big waterfalls on the Missouri, before reaching the Rocky Mountains. The perilous journey over the mountains to the Columbia River, rife with mosquitoes, blizzards and grizzly bears, drained the last of their strength. On November 7, 1805, having traveled 3,700 miles (6,000 km), Clark wrote in his journal, "Ocian [sic] in view! O! The joy."

They built a fort at the mouth of the Columbia, and hoped to find passage on a ship passing along the coast. Winter came and went and the men resigned themselves to retracing their steps homeward. They left in March 1806 and did not reach St. Louis until September, welcomed home as if from the dead.

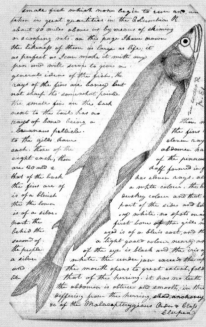

Drawing of Thaleichthys pacificus *(eulachon, hooligan or smelt) by Lewis in his notebook*

WAPATO

Lewis learned a lot about edible plants from the Blackfoot, Nez-Percé and Shoshone First Nations. They showed him how to use the potato-like tubers of the wapato (*Sagittaria latifolia*), arrowroot (*Maranta arundinacea*) for its flour, and wild ginger (*Asarum canadense*).

THE DOUGLAS FIR BEFORE DOUGLAS

In 1806 on the banks of the Columbia, Lewis described an evergreen tree 260 feet (80 m) tall, which he called the "Number 5 Fir." He drew the very characteristic scales of the cone. But it was not until 1823 that the giant tree was officially named for the great Scottish plant-hunter, David Douglas. In 1950 its scientific name became *Pseudotsuga menziesii*, after a long argument about whether it was a pine, a fir or a spruce.

"Capt. C & myself stroled out to the top of the hights in the fork of these rivers from whence we had an extensive and most enchanting view; the country in every derection around us was one vast plain in which unnumerable herds of Buffalow were seen attended by their shepperds the wolves; ... the verdure perfectly cloathed the ground ..."

Meriwether Lewis

FACING PAGE

HERBARIUM PLATE

Amorpha fruticosa

False indigo

It is very probable that this specimen was collected by Lewis and Clark themselves, but they never signed their specimens.

Amorpha fruticosa L.

ADONIS, IRIS AND GENTIAN, FLOWERS FROM THE MOUNTAINS

Augustin Pyramus de Candolle in the Pyrenées

Augustin Pyramus de Candolle was born to a notable Protestant family in Geneva in 1778. He studied medicine and botany there and then went to Paris where he met outstanding scientists such as Cuvier, Lamarck and Saussure. In 1805 he accepted a mission to survey the flora of France and he went all over the country for 6 years before becoming the head of botany at Montpellier. When the Bourbon dynasty regained power, after the 100 Days, he decided to return to Geneva. He died at 63, before finishing his Prodromus *Systematis Naturalis Regni Vegetabilis*, a masterwork that describes 58,975 species of plants. The book was completed by his son, Alphonse, and more information added by his grandson, Casimir de Candolle.

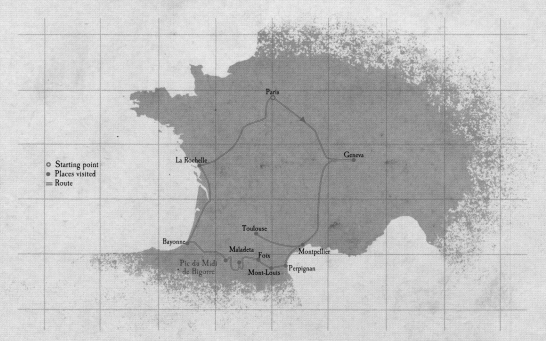

Starting point
Places visited
Route

Paris
La Rochelle
Geneva
Toulouse
Bayonne
Maladeta
Foix   Montpellier
Pic du Midi   Mont-Louis   Perpignan
de Bigorre

"On Wednesday, June 24, 1807, at two a.m., we left Collioure in the moonlight and began to enter the mountains." Walking through the night, following a guide and a pack horse, three men began crossing the Pyrenees from east to west, from the Mediterranean to the Atlantic. The expedition was led by Swiss botanist Augustin Pyramus de Candolle, whose mission was to "expand our knowledge of the indigenous plants." To do so, he verified their existence in the field or in reliable herbaria, and noted their names in local dialects. His two companions were Perrot, a naturalist, who was "cold and boring as a prison door" and Berger, "a young gardener from the Muséum and rather slow." That is how he described them to his "good little girl," his beloved wife, Fanny, who stayed in Paris. The letters he sent her were a wonderful mixture of loving words, adventures and botanical descriptions. Candolle wrote a lot; his prose was lively and modern, and his travel journal was very precise.

The expedition went from valley to valley and the botanist turned out to be a fine mountaineer. In only 2 months, he made dozens of remarkable climbs, easily going beyond the 9,800 feet (3,000 m) level on the Maladeta, the highest range in the Pyrenees, Montcalm and the Pic du Midi. The harvests were abundant and discoveries were frequent. He came back, "sinking under the weight of plants. The books, the traps, the baskets, the handkerchiefs — all were full." And there was no shortage of misadventures: storms and fog, crevices to be crossed on fragile snow bridges, lost guides and a dog eaten by two huge wolves. The scientists spent some nights sleeping outside or in shepherds' huts, eating garlic soup and sleeping on pine boughs, more comfortable than in the vermin-infested inns along their route. Their stops between Collioure and Bayonne were devoted

to drying plants. After every stage, Candolle sent the Muséum his crates of plants and any herbarium specimens he had been given, and they amounted to more than 5,000 living and dried plants. In his journal he mentioned 1,056 species and gave names to some of the newly-discovered endemic plants of the Pyrenees: hawkweed, Pyrenean buttercup, the soapwort *Saponaria caespitosa* and the lovely *Adonis pyrenaica*. When he reached Bayonne on September 1, "finished with work and fatigue," the botanist slipped away and the man wrote to his beloved, "I was quite astonished when I realized that I still had 240 leagues to travel before I could see you."

THÉORIE ÉLÉMENTAIRE DE LA BOTANIQUE, OU EXPOSITION DES PRINCIPES DE LA CLASSIFICATION NATURELLE ET DE L'ART DE DÉCRIRE ET D'ETUDIER LES VÉGÉTAUX;

PAR M. A. P. DE CANDOLLE,

Professeur de Botanique aux Facultés de Médecine et des Sciences, Directeur du Jardin des Plantes, et Membre du Conseil de l'Académie de Montpellier ; Professeur honoraire à l'Académie de Genéve ; Correspondant de l'Institut ; des Académies Royale des Sciences de Munich, Impériale de Turin , du Gard ; des Sociétés Phytographique de Gorenki , Physique de Zurich , Philomatique de Paris , de Physique et Chimie d'Arcueil, des Sciences Physiques de Genève , de l'École de Médecine de Paris , de Médecine de Marseille , d'Agriculture de la Seine et de l'Hérault, des Sciences Lettres et Arts de Montpellier , Rouen , etc. etc.

Title page, Théorie élémentaire de la botanique, ou Exposition des principes de la classification naturelle et de l'art de décrire et d'étudier les végétaux, *by Augustin Pyramus de Candolle, 1813*

FACING PAGE

HERBARIUM PLATE

Iris latifolia formerly *Iris xiphiodes*

This specimen of Pyrenees iris was collected in the Pyrenees by Candolle himself. In 1835, Candolle's herbarium contained 130,000 specimens of 75,000 species. When he moved from Montpellier to Geneva, it was transported in 40 small wagons.

"We left the next day, escorted by three guides and two mules, en route for Maladeta, a mountain some 1,760 toises high, which is the second-tallest, if not the tallest, peak in the Pyrenees, and on which no naturalist has yet set foot."

Augustin Pyramus de Candolle

Iris xyphioides. jacq.

à Pyrenæis.   Decandolle

THE GREEN TREASURES OF THE NILE

Frédéric Cailliaud, a jeweler in Sudan

Frédéric Cailliaud was born in Nantes in 1787, the son of a locksmith. He became fascinated with minerals when he was young and studied gemology in Paris. Then he traveled in Europe and Turkey, working in gem-cutting workshops and collecting rare jewels and antiques. Having earned some money in the court of the Turkish sultan, Cailliaud made his first trip to Egypt from 1815 to 1819 and a second from 1820 to 1822. He made a glorious return to France: he sold his drawings and his impressive collection of antiquities to the Cabinet des médailles of King Charles X. In 1836 he was appointed director of the Muséum d'histoire naturelle at Nantes, and lived a quiet life until he died in 1869.

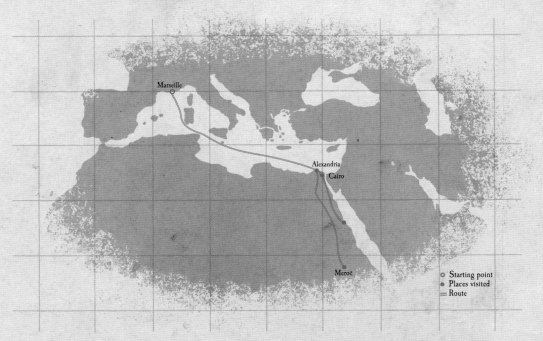

Marseille
Alexandria
Cairo
Meroë

○ Starting point
● Places visited
═ Route

On May 12, 1815, a man from Nantes who loved precious stones landed in Alexandria. Frédéric Cailliaud already had a good eye for antique objects and he was paying for his Egyptian travels with his earnings from 2 years work setting jewels in sabers for the sultan of Constantinople. The enterprising young man soon opened a jewelry store in Cairo and became friends with the French ambassador. With his new friend, Cailliaud went on an expedition up the Nile, traveling by boat, and on camels beyond the second cataract, all the way to Nubia. Cailliaud was a good observer: he described the ritual worship of the baobab, where the Egyptians burned a sheep. To pay his own "lasting homage to this divine plant," he carved his name in the bark of the huge tree.

In 1816, the Pasha of Egypt appointed him the official mineralogist, and commanded him to find the lost emerald mines of Mount Zubara, which had been lost for 2,000 years. Cailliaud set out toward the Red Sea with a few nomads, and they crossed the desert sands. He noted "very few bushes (you could walk for 2 days without seeing even one), a few spiny grasses." Almost by chance, he found the mines and had the entrance cleared. "As I sat down on the debris to rest from the labors of that day and the previous days, a fragment of dark green emerald caught my eye."

As a reward, the pasha allowed him to keep the objects he discovered in ancient tombs. When Cailliaud got back to France, he was able to sell a significant collection of papyrus, mum-

mies, jewels and statues. He returned to Egypt in 1820, this time with the official purpose of mapping the White Nile. The following summer he accompanied the pasha's son in a campaign to the south, seeking slaves and gold mines. Cailliaud followed in a barque of his own, on a crocodile-infested river that carried cargos of "bamboo, ebony, lignum vitae and other precious wood." Beyond the sixth cataract, in the heart of Sudan, Cailliaud discovered the capital of the very ancient kingdom of Meroë.

On the banks of the Nile, in deserts and at oases, the explorer collected a hundred kinds of plants, and they were given to the botanist Alire Raffenau-Delile, another "Egyptomaniac." He determined they included water lettuce, a new species of strychnine tree, various legumes including an acacia used in tanning leather and amyris, on the bark of which "the Muslims of this country write legends which they, by their custom, wrap in a small parcel of leather and fasten to their arms."

Ancient Egypt: animals in a papyrus thicket. Limestone fragment from the tomb of Neferhotep. Around 1400 BCE, 18th dynasty

THOUSAND-YEAR-OLD PAPYRUS

As he entered the tomb of Neferhotep, the "guardian of the harvest stores," at Thebes, Frédéric Cailliaud discovered amazing paintings. Among them was a scene of hunting and fishing in a vast thicket of papyrus. Papyrus has disappeared from the banks of the Nile, but once it reached 20 feet (6 m) in height and birds made their nests in its umbels. Cailliaud copied a great many paintings and then detached fragments of them with a crowbar. The papyrus thicket seen here can also be seen in the Louvre.

FACING PAGE

HERBARIUM PLATE

Ficus intermedia formerly *Ficus populifolia* Vahl

This specimen was collected by Cailliaud himself during the journey to Meroë, Sudan, and given to the botanist Raffeneau-Delile in 1826.

"Since the time of the pharaohs, perhaps, no boat had sailed on the river I was floating on ... Rough, savage nature was breathing in the middle of that constantly reborn vegetation; acacias, tiny hillocks of sand, desert dates, dead trees wound up in inextricable convolutions with vines, all forming a compact mass of greenery."

Frédéric Cailliaud

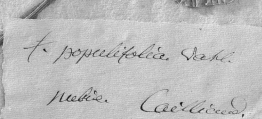

HERBIER FACULTÉ DE MÉDECINE

Herbier Delile

F. populifolia Vahl.

Nubie. Cailliaud.

HUNTING PLANTS IN
ABORIGINAL TERRITORIES
Allan Cunningham, explorer of Australia

Allan Cunningham was born in Wimbledon in 1791, to a family with Scottish roots. He first worked as a clerk but found little interest in red tape, and enthusiastically accepted a position as a secretary at the Royal Botanic Gardens, Kew. There, he learned about the flora of Australia from Robert Brown, the botanist on the Flinders expedition. In 1814 he became a collector for Kew and he sailed to Brazil and then to Australia in 1816. For 15 years Cunningham traveled the most remote regions of Australia, Tasmania and New Zealand, exploring both geography and botany. His discoveries in both domains were abundant. He died from tuberculosis at the age of 47, in 1839.

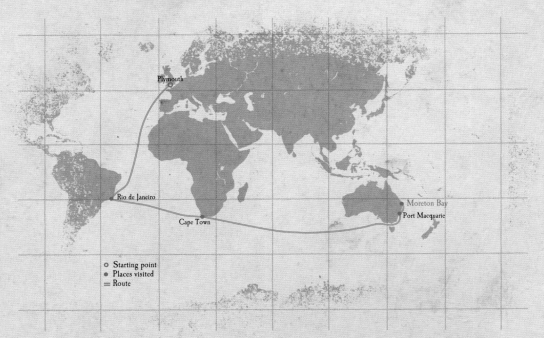

○ Starting point
● Places visited
═ Route

"So well does the serious mind of a Scottish education fit Scotsmen to the habits of industry, attention and frugality that they rarely abandon them at any time of life," stated Joseph Banks, Director of the Royal Botanic Gardens, Kew. That is why he chose Scottish botanists for the most perilous expeditions. Industrious, attentive and frugal: the very description of Allan Cunningham. At 25, he became one of the greatest plant collectors in Australia and a tireless explorer of that land. On his first expedition, in 1817, he accompanied Lt. John Oxley into the immense *terra incognita* west of the Blue Mountains. They traveled slowly along the Lachlan and Macquarie rivers, under cloudbursts that nearly sank their horses. Further on, the countryside became so arid that the men and beasts survived on short rations of water while being harassed by Aborigines. But the botanical discoveries were astounding, "Among the brushes that overshadowed the creek on which we were encamped, grew most luxuriantly, the native bignonia and a fine clematis, and being intertwined and abundantly in flower, formed the richest festoons." Even though he was very feverish, Cunningham walked almost 750 miles (1,200 km) and collected 450 specimens. From that time on, he always heeded the "call of the wild."

When he returned to Sydney, he embarked on the *Mermaid* to explore the coasts of Australia. In four voyages from the northwest to Tasmania, Cunningham discovered 700 species of plants, while dodging the stones and spears the Aborigines hurled at the invaders. While hunting plants in New South Wales, he discovered a passage toward the unexplored lands of the northwest, and decided to create his own expedition. On January 20, 1827, armed with a sextant, a pocket compass, a chronometer and a pedometer, he set out, leading six men and 11 horses, to the green prairies of modern Queensland.

In 1830, Cunningham went to spend several years in England, while his youngest brother, Richard, arrived in Australia. He also was a brilliant botanist but got lost in the bush while hunting plants. He was found and fed by the Aborigines but in his delirious condition, suffering from thirst and fear, he could have given away their hiding place, so, having considered the matter, the Aborigines killed him. Allan Cunningham, although deeply affected, went on to New Zealand to search for plants. There, he caught tuberculosis. In his final letter, he described himself as a "poor traveler, decrepit and prematurely aged," who, "purely out of love," combined "the study of the country's plants with that of its geography."

GREEN PRAIRIES

Cunningham considered the Darling Downs his foremost discovery. Where Brisbane now sits, he found green prairies where "the native grasses had resisted the dry weather ... They were fresh, verdant, and doubtless nutritive, affording abundance of provision to the many kangaroos that were bounding around us."

A UNIQUE CHESTNUT TREE

The *Castanospermum australe* A.Cunn., called the Moreton Bay Chestnut or Black bean tree, is the only chestnut-like tree in Oceania. Cunningham discovered it around Moreton Bay. Its seeds were traditionally consumed by Aborigines and are now being studied for their potential to inhibit the AIDS virus.

Map of the Brisbane River drawn by Allan Cunningham

"Another sudden change happened in the instant when we left the mire and its stunted eucalyptus and entered the deep shade of the tropical forest, composed of the massive boles of that red cedar, the turpentine tree or Tristania albens ... and Alsophila, a tree-like fern of New South Wales, all tied together by immense twining plants that even the least interested observer would have to notice and admire."

Allan Cunningham

FACING PAGE

HERBARIUM PLATE

Araucaria cunninghamii

The specimen of hoop pine in the middle was collected in Australia in 1818 by Cunningham himself.

Araucaria cunninghamii D·Don

Det. A. Farjon (RBG Kew) Sep 2004

LECTOTYPE of *A. cunninghamii*
D·Don, in Lambert, Descr. Genus
Pinus, ed. 2, vol. 3, Sub t. 102. 1837
SPECIMEN ON FAR LEFT ONLY
Det. A. Farjon (RBG Kew) Sep 2004

Araucaria cunninghamii Aiton ex D. Don
var. *cunninghamii*

Det. A. Farjon (RBG Kew) Dec 2006

ALLAN CUNNINGHAM'S
AUSTRALIAN HERBARIUM.

Presented by Robert Heward Esq. 1862.

BR
19091

KEW HERBARIUM

0000478

R 20049

Araucaria Cunninghamii
Aiton

East coast
New Holland
1818 – 1829
Cunningham

KEW HERBARIUM

0000477

London Exhib. 1862

No 2 Queensland woods

Araucaria Cunninghamii
Ait.

Mr Hill,

*Araucaria
Cunninghamii*
Coult. Bevijbl

IN LOVE WITH THE HYDRANGEA
Philipp Franz von Siebold in Japan

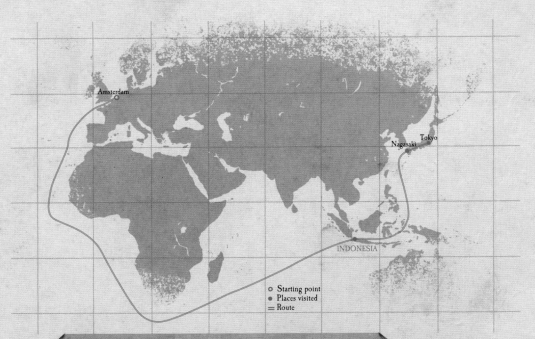

Philipp Franz Balthasar von Siebold was born in Bavaria in 1796. His father, grandfather and uncles were professors of medicine at the University of Würzburg, where he began his studies in 1815. In 1822, he became a doctor in the army of the Dutch East Indies and was stationed at Batavia (Jakarta). He was chosen to go to Japan because of his serious attitude and his talent for languages. He was expelled from Japan in 1829 and settled in Leiden, the Netherlands, and worked on his immense naturalist collections and his account of the voyage to the court in Tokyo for the next 20 years. He returned briefly to Japan as a diplomat, from 1859 to 1863, but died in Munich in 1866.

○ Starting point
● Places visited
═ Route

In 1823, Japan was still firmly closed to foreigners. Two Dutch ships per year were authorized to land at Dejima Island in Nagasaki Bay, and foreigners were not allowed to leave the islands. But one day Doctor Siebold was called to the mainland to look after an important Japanese man. His success with that patient enabled him to open a small clinic and to meet with advanced students of Western medicine who spoke Dutch. Then love entered the picture; it was so strong that Kusumoto Taki, whom he called O-Taki-san, accepted the shameful status of concubine because she could not marry a foreigner. Soon, they had a little girl, Kusumoto Ine, also known as O-Ine. Following the Siebold family tradition, she went into the field of medicine and eventually became the first woman doctor in Japan.

Siebold gradually enlarged his medical practice and continued to give courses. As it was forbidden to pay him in money, his patients gave him gifts, and thus the doctor amassed thousands of everyday objects, books and prints, which he sent to Batavia, Brussels and Antwerp. He had his students bring him seed and plants, and smuggled tea plants to Java, where they flourished. His collection became so vast that two assistants, an apothecary and a painter, were sent to help him.

In 1828 his reputation had reached the Imperial court and he was invited by the Shogun to Edo (now Tokyo). He came back loaded with plants, animal car-

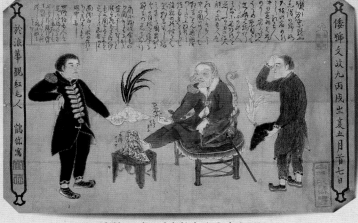
Siebold attending a kabuki play in Osaka in 1826

casses and *objets d'art*, and also with a map of Japan, which was a state secret at the time. Siebold hastened to send this precious cargo to Europe but the boat sank and the Japanese authorities found the contraband map. The punishment was severe; Siebold's students were imprisoned and he was banished forever. His companion and their daughter were not allowed to go with him.

Back in Holland, Siebold threw himself into his work, sorting and describing his fabulous collections. He only managed to study 2,300 of the 12,000 specimens but was able to add to the work of the pioneer of Japanese botany, Carl Peter Thunberg. He also had greenhouses built from more than 1,000 trees spread across Europe. Among them are the Japanese cherry tree with its spectacular pink flowers, the fragrant Siebold's or Oyam magnolia (*Magnolia sieboldii*), *Ginko biloba* with its golden leaves, as well as hostas, azaleas and hydrangeas. Siebold named the luminous blue *Hydrangea otakas* for his beloved companion (it is now called *Hydrangea macrophylla*).

MINIATURE GARDENS

Siebold was fascinated by the miniature landscapes created by Japanese gardeners, with rocks, mountains, lakes and forests in pots. The adult bonsai trees were no more than 3 feet (1 m tall) and had "luxuriant branches on a distorted trunk."

JAPANESE INVADERS

Descendants of Siebold's few plants of Japanese knotweed (*Fallopia japonica*) have now become invasive in Europe and North America. Knotweed's dense, spreading rhizomes make it resistant to drought and flood, and it crowds out indigenous plants.

FACING PAGE

HERBARIUM PLATE

Rubus hirsutus formerly *Rubus thunbergii*

This specimen was collected in Japan by Siebold himself and mounted in 1843.

"Around the temples and monasteries in these wild lands, the ordered mind has created bewitching copses dressed with multicolored azaleas, camellias, peonies, and superb lilies and orchids."

Phillip Franz von Siebold

3

Herbarium Zuccarinii.

Legi in Japonia de Siebold

Communicavit *Zuccarini* anno 18

China Fortune 1846

R. Thunbergii, S. & Z.
Rauini.

INDEX FLORÆ SINENSIS.

HERBARIUM 1854. BENTHAMIANUM

HERBARIUM 1854. BENTHAMIANUM

Herbarium Zuccarinii.

Rubus

BENEATH THE PACIFIC ALGAE
Jules Dumont d'Urville in Oceania

Jules Sébastien César Dumont d'Urville was born into a noble family in Normandy in 1790 and entered the French navy at 17. His first long expedition was a voyage around the world on *La Coquille* with Duperry from 1822 to 1825. Dumont d'Urville was then named captain of *L'Astrolabe* and sailed the Pacific from 1826 to 1829, but fell out of favor during the Restoration. Finally, in 1837 he went on a great journey of exploration towards the South Pole and was the first to find land in Antarctica. He named it Adélie Land, after his wife. He, his wife and young son died in France's first railway disaster in Meudon in 1842.

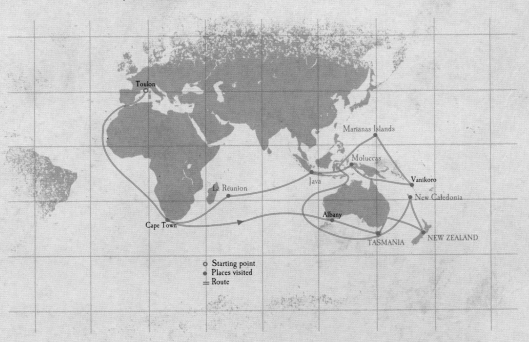

○ Starting point
● Places visited
═ Route

It was an Irish seal hunter who gave France the news about la Pérouse who had vanished 38 years previously in the South Pacific. In 1826 Peter Dillon bought a French sword hilt and a Saint Louis cross from natives of Vanikoro, in the Solomon Islands. Immediately, an expedition set out from Toulon, commanded by Jules Dumont d'Urville who had just completed a successful voyage around the world on board *La Coquille*. The botanist on board that ship had been René Primevère Lesson, whose claim to fame is that he was the first naturalist to see hummingbirds and birds of paradise. In addition he brought back more than 3,000 species of plants to the Muséum.

For its second voyage, *La Coquille* was renamed *L'Astrolabe*. The ship's botanist, a naval surgeon, was Pierre Adolphe Lesson, younger brother of René. As a fervent naturalist himself, Dumont d'Urville made several long stopovers to collect specimens. The most productive was in New Zealand, whose isolation had encouraged development of an abundant endemic flora that had not yet been discovered. The naturalists carried on enthusiastically, despite attacks from cannibal Maoris. They were also intrigued by the algae found in these warm waters. Pierre Adolphe Lesson, who was not fond of his captain, named a spiny giant kelp *Marginaria urvilliana*.

In February 1838, the expedition finally arrived at Vanikoro. Dumont d'Urville anchored *L'Astrolabe* very

carefully; the little island is surrounded by a coral barrier reef with many false passages that were literally dead ends for large vessels. The officers began asking the natives about what happened to La Pérouse, but everyone seemed to have amnesia. Dumont d'Urville despaired, and both he and his men were suffering greatly from fever. The second mate of *L'Astrolabe* had the idea of "offering a piece of red cloth" for which one reticent islander agreed to take them to the site of the shipwreck. In the transparent lagoon lay the key to the La Pérouse mystery. "The native stopped the canoe and indicated that the Frenchmen should look into the water. Twelve or 15 feet below, they saw many objects scattered about and covered with coral, including anchors, cannons and cannonballs." Dumont d'Urville raised these vestiges, erected a wooden monument to the unfortunate explorers, and left Vanikoro in haste. "Our future looked grim if we could not soon leave these dismal shores."

THE LESSON BROTHERS

René Primevère and Pierre Adolphe Lesson were both naval surgeons and naturalists, and both had wide-ranging curiosity. They left their native city of Rochefort many written records, dictionaries of Oceanic languages, ethnographic objects, stuffed animals and three mummified Maori heads.

VANIKORO 2005

In recent decades, seven scientific expeditions have tried to discover more about La Pérouse's shipweck. Finally an expedition called Vanikoro 2005 identified the wreck of *La Boussole* sunk into a false gap in the reef, while La Pérouse's *L'Astrolabe* was found by Dumon d'Urville in a breach of the coral barrier.

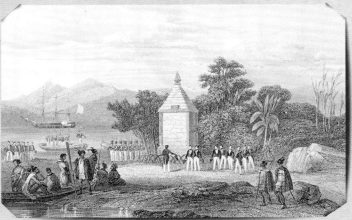

Monument to the memory of Jean-François Galaup, comte de La Pérouse, erected by Dumont d'Urville. 19th-century engraving

"On the hillside and in among many other species, they showed me the kauri, which produces the best wood in New Zealand, in the opinion of both the natives and the missionaries. The latter employ it in all kinds of construction, and the former make their best dugout war canoes of it."

Jules Sébastien César Dumont d'Urville

Eucaliptus
angustifolius
5 Dec
non Spott

THE BUSH,
A SECRET GARDEN
Georgiana Molloy, an Englishwoman in Australia

Georgiana Kennedy was born to a middle-class English family in 1805. She was a pious child, which distanced her from her parents, and she went to live with a friend. In 1829, after much hesitation, she agreed to marry Captain John Molloy and immediately sailed with him to the British colony of Western Australia. Pioneer life was very difficult and she lost two children very young. Captain James Mangles urged her to send her wonderful collections of plants and seeds to England, and she began to do so in 1836. Worn out by many pregnancies and her incessant work collecting plants, Georgiana Molloy died at 37, in 1843.

Portsmouth

Cape Town   Augusta

○ Starting point
● Places visited
═ Route

When she married Captain John Molloy the blonde, delicate, cultivated Georgiana knew that she would have to go with him to a distant colony in Western Australia, but she did not have a real understanding of the harsh life of a pioneer woman. When she arrived at Augusta in March 1830, she was 6 months pregnant and her only lodging was a tent on the sandy shores of the Swan River. Weakened by the journey, she gave birth to a baby who did not survive. She buried the baby and covered the grave with English clover.

Her husband was away working for long months at a time and the young woman suffered from loneliness among the rough-hewn settlers. "Cattle, swine, whale hunting, harpoons, potatoes and onions are the main subject of conversation," she wrote to her friend Helen. Then there were other children and Georgiana was overwhelmed with the household chores. To keep track of her youngest boy she attached a bell to his belt. One day the bell was silent; he had drowned in the well.

Happily, in 1836 a letter came from England. James Mangles, a retired sea captain and amateur botanist, offered to exchange plants and seeds with Georgiana. She threw herself into this project and began to tame the savage Australian bush, little by little. She sent Mangles many dried plants, arranged tastefully and with precise information on how they grew and how they were traditionally used by the Aborigines. Her meticulously prepared seeds germinated successfully in prestigious European nurseries. Mangles was shamelessly selling the Molloy collections and taking credit himself. Any new species were described by Professor John Lindley who became the official discoverer. Georgiana developed a passion for botany. His descriptions showed more sensuality than a Victorian education would predict: "I discovered a plant that took my breath away, a little white, delicate flower, on a bush resembling gorse."

Her health was declining, but she still had the strength to collect seeds of the dazzling orange-flowered *Nuytsia*. "The seeds of *Kingia* and *Nuytsia* you shall have," she wrote to Mangles. "This is their precise time of ripening and the last named grows here in great abundance, and splendid it is. It is so rich among all the sombre Eucalypts of the present season. It presents to my mind the rich and luxurious tress which adorn paradise ..." In her final letter she wrote, "I think I have sent you everything that deserved to be sent," and died after giving birth to her seventh child.

TRUMPETS OF FAME

Georgiana Molloy's discoveries were published in 1839 in *A Sketch of the Vegetation of the Swan River Colony* by Professor Lindley of University College, London. Among the dozens of new plants were the eucalyptus with ruffled flowers (*Corymbia calophylla*), orchids, carnivorous sundews (*Drosera*) and acacias, but the only plant named for this unknown botanist is a shrub 10 feet (3 m) tall, with fragrant pink bell flowers. In 1842, it was proposed that this plant be named *Boronia molloyi* (the Latin ending indicating a man's name), then changed to *Boronia elatior* in 1844, and finally, in 1970, it officially became *Boronia molloyae*, with a feminine ending.

Reproduced by permission of Mrs A

Fair Lawn, Georgiana Molloy is on the horse on the right.

"I saw a tree of great beauty, dark green and thorny. The flowers are of purest white and fall from the stems in long tresses. Some buds were floating in the breeze like snowflakes and sending out the most delicate of perfumes, like the bitter almond."

Georgiana Molloy

═══ FACING PAGE ═══

HERBARIUM PLATE

Caladenia flava

These orchid specimens were collected near the Swan River in Australia by Georgiana Molloy herself in 1839. The herbarium voucher unfairly gave credit to James Mangles.

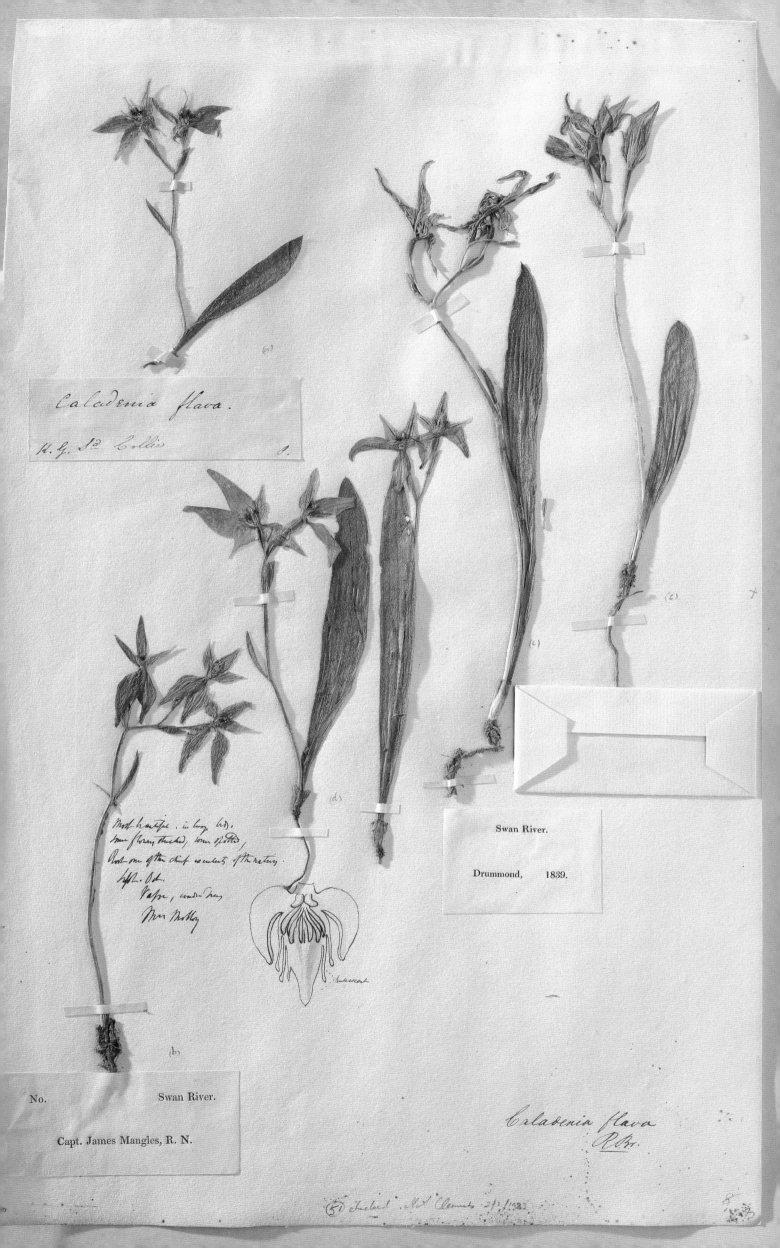

Caladenia flava.

H. G. S³ Collie

Swan River.

Drummond, 1839.

No. Swan River.

Capt. James Mangles, R. N.

Caladenia flava
R.Br.

THE MYSTERY OF MYSTERIES
Charles Darwin in the Galapagos

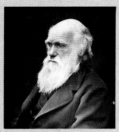

Charles Darwin was born in 1809 in Shrewsbury, England; his family was rich in poets, doctors and zoologists. He studied medicine at Edinburgh and humanities and geology at Cambridge. His only enthusiasm was his beetle collection. For the rest of his life, after traveling around the world on the *Beagle*, Darwin suffered nausea and vertigo. That did not prevent him from marrying and having 10 children. Living in a little village in Kent, he made use of the vast amount of material he had brought home from his travels. He published his travel journal in 1839, and many other works, including the well-known *On the Origin of Species* in 1859 and *The Descent of Man* in 1871. Darwin died in 1882.

Plymouth
Azores
Cape Verde
Galapagos
Bahia
Rio de Janeiro
Valparaiso
Montevideo
Falkland Islands
Cocos (Keeling)
Mauritius
Sydney

○ Starting point
● Places visited
═ Route

Perhaps no one has influenced our knowledge of life on earth as much as the young geologist who embarked on the *Beagle* at Plymouth that cold December morning in 1831. He was only there because the more experienced naturalist has resigned. The *Beagle*'s captain, Robert FitzRoy, a believer in phrenology, nearly rejected Darwin because the shape of his nose did not indicate a propensity for long voyages. But the young man was a gentleman, which means that he had received an excellent education and did not need to be paid for his work. Those were the conditions the captain imposed as he prepared to share his personal cabin for 2 years. In fact, the voyage lasted 5 years and the conservative, pro-slavery Fitz-Roy's patience would sometimes be tested to the limits by his liberal, abolitionist passenger. Darwin noted in his journal, "The difficulty of living amicably with a captain of a warship is considerably increased when it amounts almost to mutiny to speak to him as one would speak to any other man."

Once they were out to sea, the young naturalist suffered enormously from seasickness, which was to torment him often. But at the first port of call, he began collecting things that, like pieces of a jigsaw puzzle, would become his revolutionary theory on species and the origins of man. In Cape Verde, Darwin wondered about the presence of fossil shells in a cliff, some 65 feet (20 m) above sea level. Little by little, geology began to reveal the unfathomed depths of time.

In February 1832, the *Beagle* made stops in South America, first at Bahia, then at Rio. Darwin plunged into the exuberant nature of the tropics: "The ele-

Engraving of a Galapagos tortoise (Testudo aningdonii). Illustration from the first illustrated edition of The Voyage of the Beagle: Charles Darwin's Journal of Researches, *1890*

gance of the grasses, the novelty of the parasitical plants, the beauty of the flowers, the glossy green of the foliage, but above all the general luxuriance of the vegetation, filled me with admiration." He measured, took notes and collected as many rocks, plants (rubber tree, trumpet vine and the giant flower of *Aristolochia gigantea*) and as many as 68 different beetles in a day.

Farther south, in Patagonia and Tierra del Fuego, his encounter with the natives wearing animal skins was fixed in his memory and convinced him that every human being, even the most civilized gentleman, has an animal component.

On the Argentine coast he mingled with the gauchos and those cowboys of the vast pampas quickly adopted "Don Carlos," who was a good horseman and so curious about their understanding of their environment. Darwin observed that some species (ostriches, for example) changed as one traveled toward the southern reaches of the continent. Why? Near Montevideo,

(continued on page 108)

BEAUTY AND SURVIVAL

In the early 19th century, the idea of natural selection had many detractors. Many people were convinced, for example, that the beauty of flowers was created to satisfy the human eye and not to ensure the survival of the plant. Darwin decided to prove the opposite, that each whorl in a petal and every fragrance had a specific function. To do so, he chose to study the enigmatic reproductive process of orchids. He discovered that it was the work of 26 species of moths and proved that the flower had the perfect structure to provide these nocturnal visitors with their packets of pollen.

FACING PAGE

HERBARIUM PLATE

Berberis darwinii

The specimen in the middle of the page was collected by Darwin himself in Chile. At that time, a number of specimens from the same species but coming from different places and different collectors would be mounted together to save paper.

"Bahia, or San Salvador. Brazil, Feb. 29th.—The day has past delightfully. Delight itself, however, is a weak term to express the feelings of a naturalist who, for the first time, has wandered by himself in a Brazilian forest."

Charles Darwin

Bridges.

585. Berberis
In los Llanos, between
Valdivia & Osorno,
3 3-6 f.t

Valdivia
Bridges

Berberis Darwinii Hook. & Arn.

Determinavit   Maria M. Job
                        10-1-1939

Berb. Darwinii. Hook.
Ic. Plat. 672.

7/1-1-369 TYPE

he discovered the fossils of giant quadrupeds, Megatherium, Macrauchenia and Toxodon. He noticed that some of them strangely resembled armadillos and sloths. If some species had disappeared, how did new ones appear? "Mystery of mysteries"... dare he think that they were not all created by God?

Before the *Beagle* traversed the Strait of Magellan in June 1834, Darwin sent an impressive collection back to England. Visiting Chile, Peru and the islands had provided the young man with frightening experiences such as earthquakes, tsunamis and volcanic eruptions, and enlightening observations of geology, flora and fauna. He discovered *Gunnera*, a giant wild rhubarb, red cedar, Patagonian cypress, fields of fuchsias and much more.

It was in the Galapagos Islands that Darwin, unaware, had the encounter that would change his life. In September and October 1835, while the *Beagle* was sounding the depths, Darwin explored four tiny volcanic islands where iguanas and tortoises lived. "These huge reptiles, surrounded by the black lava, the leafless shrubs and large cacti, seemed to my fancy like some antediluvian animals. The few dull-colored birds cared no more for me than they did for the great tortoises." Those insignificant birds, however, would eventually set off a big bang in natural history. On the four islands, Darwin trapped a dozen of these finches with differing beaks, as well as 193 plants, including 100 endemic new flowers.

The return journey took them 3,100 miles (5,000 km) across the Pacific to Tahiti, where Darwin studied the coral in the lagoons, to New Zealand, Australia, Mauritius and South Africa. At Ascension Island he received a letter from his sister reporting that his work, as known from his letters and shipments, already promised him a "a place among the great men of science." On his return in October 1836, Darwin began to work on his discoveries. An ornithologist explained to him that "his" finches belonged to different species, each one particular to a single island. He then understood that isolation, diet and climate could lead to the formation of distinct species from a common ancestor.

From finches to monkeys to man, it was only a small step, but such a daring step that Darwin hesitated for 20 years before proposing the idea to the scientific community in 1859.

Drawing of Galapagos finches. Charles Darwin observed a number of species of finches whose only differences were in their beaks, each beak being adapted to a particular type of food.

CARNIVOROUS PLANTS

Darwin was fascinated by carnivorous plants that lived on poor soil and fed on insects. Trying to trigger the digestive process of a *Drosera*, he tested milk, meat, hair and glass, but only animal-based substances interested the *Drosera* and Darwin, amazed, discovered that it secreted real digestive juices.

THE POWER OF MOVEMENT

As he aged and his health declined, Darwin became more interested in the plants he could observe at home. He said he was "fascinated and perplexed" by climbing plants. He was able to draw the movement of their tendrils as they sought support. And he concluded that, if necessary, these plants could acquire the power to move around.

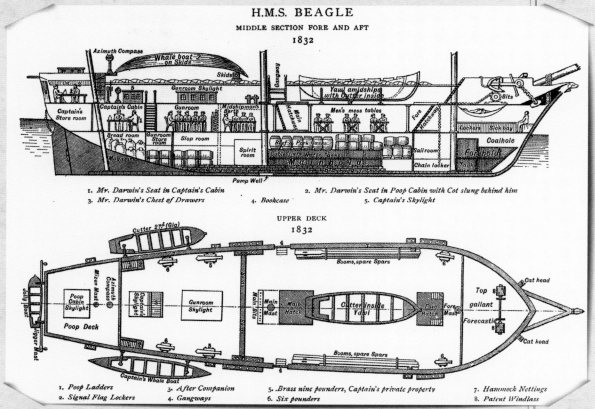

Plan of the H.M.S. Beagle on which Darwin sailed

"*Considering the small size of these islands, we feel the more astonished at the number of their aboriginal beings, and at their confined range. Seeing every height crowned with its crater, and the boundaries of most of the lava-streams still distinct, we are led to believe that within a period, geologically recent, the unbroken ocean was here spread out. Hence, both in space and time, we seem to be brought somewhat near to that great fact — that mystery of mysteries — the first appearance of new beings on this earth.*"

Charles Darwin

FACING PAGE

HERBARIUM PLATE

Adiantum henslovii

This specimen was collected in the Galapagos by Darwin himself in 1835.

Adiantum Henslovis & Jos. Hooker MS.
Galapagos; S. Amer:
(Charles Island)
Sept. 1835. C. Darwin.

H·346/89
29
KEW

photo - GH
TRYON 1960

MEGAHERBS AND RHODODENDRONS

Joseph Dalton Hooker, from the South Pole to the Himalayas

Joseph Dalton Hooker was born in 1817, and became a botanist almost immediately. His father William Hooker was an eminent professor at the University of Glasgow and young Joseph attended his courses when he was barely 5. Joseph studied medicine, which led in those days to botany. From 1839 to 1843 he joined James Clark Ross's expedition to the Antarctic aboard the *Erebus* and *Terror*. From 1848 to 1851 he went to northern India and Nepal; in 1860 he was in Palestine, in 1871 in Morocco, and in 1877 in the United States. A great friend of Darwin, Joseph Hooker succeeded his father William as head of Kew Gardens in 1865. He supervised Kew's great expansion before his retirement in 1885 and died in 1911.

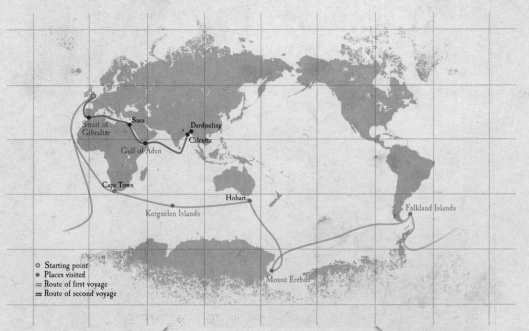

Strait of Gibraltar · Suez · Dardjeeling · Calcutta · Gulf of Aden · Cape Town · Hobart · Kerguelen Islands · Falkland Islands · Mount Erebus

○ Starting point
● Places visited
═ Route of first voyage
═ Route of second voyage

"I thought I should be the happiest boy alive if ever I would see that wonderful arched rock, and knock penguins on the head." That was the strange childhood dream of Joseph Dalton Hooker, who was growing up on the tales of his grandfather's travels and the stories of Captain Cook. The Antarctic penguins were no doubt surprised to meet Joseph when he was only 22. A graduate of medical school, young Hooker was invited by James Clark Ross, an explorer and friend of his father, to take part in an exceptional scientific expedition to the Antarctic. He was officially the ship's assistant surgeon and botanist. He was somewhat nervous but full of ambition, because he was following the trail of Darwin on the *Beagle*. He wrote to his father that Darwin, "I daresay, knew his subject better than I now do, but did the world know him? The voyage with FitzRoy was the making of him (as I had hoped this expedition would me)."

The *Erebus* and *Terror* left England in September 1839; they were refitted bomb vessels, their hulls reinforced with metal to resist the pack ice. At Madeira, at Cape Town, at sea, Hooker worked ceaselessly. He impressed Captain Ross so much that he allowed him a space on his own desk to set up a microscope and more space to store his large collection of plants. To ward off chances of misadventure at sea, the young botanist collected one specimen for the Royal Society, one for the Admiralty, two for his father at Kew and one for Captain Ross. On the Kerguelen Islands, where Cook had seen only 20 plants, Hooker identified 150 different species, including 18 flowering plants, 3 ferns and 35 mosses, lichens and algae. And that was in a climate where samples had to be taken with a hammer and the lichens warmed with body heat.

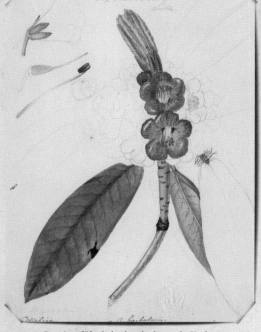

Drawing of Rhododendron barbatum by Hooker

In 1840, on New Zealand's sub-Antarctic islands, Hooker made his great discovery: tall flowers in bright yellow, mauve and pink, with enormous leaves. Their size was all the more surprising because most vegetation in the extreme south is dwarfish. Ross named them the "megaherbs" and today they are known as *Pleurophyllum hookeri*, *Bulbinella rossii* and *Anisotome latifolia*, among others.

In January 1841, *Erebus* and *Terror* reached their southernmost point, at 78°3' S; Ross was able to locate the south magnetic pole and Hooker to knock a penguin on the head. The crew, fascinated, watched the eruption of Mount Erebus and then sailed for Tasmania before launching a new incursion into the

(continued on page 112)

RENAISSANCE OF THE MEGAHERBS

In the 1800s many of the Antarctic islands were stocked with rabbits and goats, in order to help shipwrecked English sailors survive, but the animals cleaned off the native vegetation. In 1993 a New Zealand conservation program removed the invasive animals and the lost megaherbs returned, in great abundance and to everyone's surprise.

DOCTOR WARD'S INVENTION

Around 1829, Doctor Nathanial Ward was growing his ferns in sealed, glass and wood boxes to shelter them from London's air pollution. He had Australian plants shipped to him in these cases and they arrived in flourishing condition after a 2-year voyage. Joseph Hooker was the first to use these terrariums, known as Wardian cases, thus beginning an advance in techno-botany that had immense economic consequences.

"... and all the coast one mass of beautiful peaks of snow, and when the sun gets low they reflect the most brilliant tints of gold and yellow and scarlet, and then to see the dark cloud of smoke tinged with flame rising from the Volcano in one column, one side jet black and the other reflecting the colors of the sun ... extending many miles to leeward."

Joseph Dalton Hooker

═ FACING PAGE ═

HERBARIUM PLATE

Bulbinella rossii, formerly *Chrysobactum rossii*

This specimen of a megaherb was collected on Campbell Island by Hooker himself in 1845.

Chrysobactrum
Nepii Hook fil
♂
Tamphledes Ho
Jos. Hooker 1845

icy seas that lasted all of 1842. On the Falkland Islands, Hooker discovered many plants that closely resembled British species and began his lifelong investigation of the spread of species. When he returned to England, these ideas were the basis of a long friendship with Darwin.

By 1848, the young man had earned fame as a botanist. His father obtained support for him to explore the north of India and collect plants for Kew Gardens. Joseph Hooker landed at Calcutta, then went on to Darjeeling where he met a British government official, Archibald Campbell. Together they went on an an expedition to Sikkim, a small state surrounded by Tibet, Nepal, Bhutan and British India. The rajah was weak and his prime minister, the dewan, wanted to protect them from the English. He reluctantly permitted this expedition but forbade them to cross the border into Nepal.

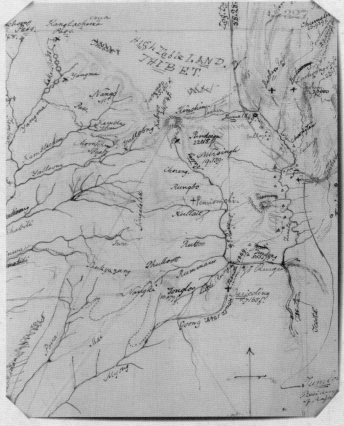

Map showing two access routes to Darjeeling through the Tibetan passes, drawn by Joseph Dalton Hooker

KEW GARDENS AND THE HOOKER FAMILY

In 1841, Sir William Hooker, father of Joseph, was appointed Director of the Royal Gardens at Kew, which had been neglected since the death of Sir Joseph Banks. He wanted to make it the primary center of scientific botany in the British Empire, particularly in the study of plants of economic interest (tea, opium, textile-producing plants, etc.) and their cultivation and processing. He constructed the large tropical greenhouse, the herbarium and the library, and he opened the gardens to the public. His son succeeded him in 1865 and Kew Gardens grew from 11 acres to 300 in 20 years, adding more than 20 greenhouses and 4,500 plants, including rubber trees and cinchona.

Hooker and Campbell launched their assault on the Himalayas with 56 porters — 7 just to carry plant specimens — plant and bird collectors, and guards. Hooker himself carried a small barometer, a large knife, a spade, a notebook and a telescope. He was excited by the tropical forest and its two flowering seasons, one in summer and one in winter, saying, "I know nothing more beautiful than a branch of *Rhododendron argentum*, with its spread out foliage and its glorious masses of flowers," and "The leaves in this season are brilliant green and as the evening advanced, a yellow convulvus erupted into bloom, as if by magic, adding more beauty to the bushes on which it climbed." The explorers climbed the Himalayan foothills above 10,000 feet (3,000 m), met Tibetans who had never before seen white men,

and slept in yak-skin tents. Hooker collected high-altitude plants (dwarf bamboo, barberry, saxifrage and lichen), and despite the dewan's ban, the two Britons crossed into Nepal. In November 1849 they were arrested and jailed, and it took the threat of invasion to get them out of the prisons of Sikkim.

The expedition continued to Bengal, by elephant and by boat, always searching for plants. By 1851 Hooker had collected some 7,000 species, including 25 new species of spectacular rhododendrons. The magnificent illustrations of these plants by Walter Hood Fitch in Hooker's first book, *The Rhododendrons of Sikkim-Himalaya*, set off a fever in British gardeners and a mania for rhododendrons. They still bloom in English gardens.

Drawing of the Tibetan landscape from Hooker's notebooks

"The summit of the pass was wide, grassy and covered with bushes of dwarf bamboo, roses and barberry covered with mosses and lichens. It had rained all morning and the plants were covered in ice ... I collected some very curious and beautiful mosses, placing these icy treasures into my box as if they were exquisite glass ornaments."

Joseph Dalton Hooker

FACING PAGE

HERBARIUM PLATE

Rhododendron dalhousiae

This rhododendron specimen was collected in India by Hooker himself in 1850.

Rh. Dalhousiae, Hook. f.
(ex Hook. Ic. orig.! Fitch partim)
Determinavit J Hutchinson.

JDH 1850. TYPE
R. dalhousiae
seeds.

J.D.H.
1850

FLORA BRIT. IND. Vol. III. 469
Named by Mr. C. B. CLARKE.

Rhod. Dalhousiae JH.
Darjiling 7000ft.

JDH

HERBARIUM
1867
HOOKERIANUM

CARGOS OF FLOWERS FROM THE INDIAN OCEAN

Mathurin Jean Armange, Master Mariner

Mathurin Jean Armange was born to a family of sailors in 1801 in Brittany. He joined the merchant marine as a cabin boy at the age of 11. Over numerous voyages between Nantes, la Réunion and India, he rose through the ranks to become a master mariner. From 1841 to 1863, Armange, who was fascinated with botany, brought many seeds and live plants back for the Société nantaise d'horticulture, which consisted of a number of wealthy and educated amateurs, and for the botanical gardens of the Muséums of Paris, Nantes, Le Havre and Rochefort. He died in 1877 at Nantes.

Jean-Marie Écorchard, Director of the Nantes botanical garden

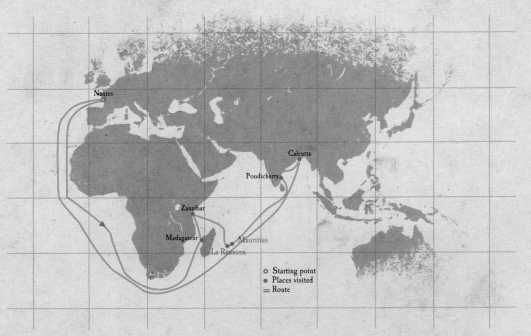

- ○ Starting point
- ● Places visited
- ▬ Route

"There is one navigator, gentlemen, whose name is mentioned in every report, and that is Captain Armange." The Société nantaise d'horticulture could hardly contain its enthusiastic praise for this "friend of botany," "intrepid navigator," "courageous and unselfish importer," and generous donor of rare plants from the Indian Ocean. In 1842 he "brought 13 crates containing 664 plants or shrubs, and more than 2,000 samples of dried plants from Bourbon or India," and in 1843, "more than 200 new plants were provided to the botanical garden." Their health and freshness after "more than one hundred days of sailing attested to Captain Armange's intelligent care." The Captain was skilled in the use of terrariums.

Like a dozen other colleagues, Mathurin Jean Armange was continuing a long relationship between Nantes and exotic flora. As early as 1719, because of its mild winters and location on the Loire, the port of Nantes had been chosen for the royal botanical garden. It was the first stop on French soil for plants that came from the four corners of the world. The plants that made it through the cold, the rolling seas, salt spray and lack of fresh water were revived there before being sent to the Jardin du roi in Paris. Louis XV even issued an edict obliging "Captains and masters of ships of Nantes, to bring back with them seeds and grains from colonies in foreign lands."

At the helm of *Anna*, Armange engaged in commerce in the Indian Ocean, exchanging manufactured goods for sugar. Beyond Zanzibar, Madagascar, l'île Bourbon (la Réunion) and Ceylon, he traveled to the coasts of Coromandel and Bengal. It was a long journey, plagued by monsoons and "all the miseries suffered by those who devote themselves in all climates to the dangers of forests and perils of the seas." Still, for 20 years, Armange filled his hold with baskets of bulbs, crates of seeds and thousands of plants: fruit trees, ferns, orchids, palm trees, amaryllis, water lilies, flamboyant trees, hibiscus, dracaena and more. He not only enriched French collections, he also transported many species from one colony to another, from Calcutta to la Réunion, from Pondicherry to Mauritius. His zeal won him the gold and silver medals of the Société nantaise d'horticulture and this testimonial: "Armange is one of those unfortunately rare men who know how to devote the scarce free time of a professional mariner to studying the natural sciences for the betterment of their country."

THE DEADLY EQUATOR

In his correspondence with the Muséum d'histoire naturelle in Paris, Captain Armange regretted the frequent loss of tropical plants when crossing "the line." "As we approached 5° or 8° N, within three or four days I had lost a great many of the plants I intended for you."

PORT GARDENS

Port gardens were created in Lorient, Brest, Nantes, Rochefort and Toulon. These were small, enclosed gardens with orangeries and heated greenhouses. The tropical plants were "comforted," "refreshed," and "healed" before continuing their journey to Paris by cart or barge.

List of plants from the Indies and Bourbon contained in Captain Armange's crates on his return to Nantes, 1849

"Among these plants, there is one that is particularly useful on the island of Bourbon, and that is a banana tree, with vigorous growth and an abundance of excellent fruit, which has replaced the former species ... the people there call it the Armange banana; I carried that plant from Bourbon to Mauritius."

Mathurin Jean Armange

FACING PAGE

HERBARIUM PLATE

Acanthus ilicifolius L. formerly *Dilivaria ilicifolia*

This acanthus specimen was collected in Bombay, India.

Mijin Madden

Bombay
Delvaria

Bombay
W. Law

Mijin Madden

19

Dilivaria
ilicifolia
Pointt de Galle, March 18.

HERBARIUM
1867.
HOOKERIANUM

HERBARIUM
1867.
HOOKERIANUM

HERBARIUM
1867.
HOOKERIANUM

CARNIVORES AND ORCHIDS IN THE LAND OF THE WHITE RAJAH

Hugh Low in Sarawak

Hugh Low was born near London, England, in 1824, just after his father had opened a nursery for greenhouse plants. Young Hugh loved plants and was a careful observer. At barely 17, he was hired by the East India Company and left for Borneo. In 1844 he met James Brooke, a former British officer who had become the Rajah of Sarawak in Malaysia. As Brooke's private secretary he had the opportunity to explore the region's forests. Later, Low spent 40 years as the administrator of the provinces now included in the sultanate of Brunei, and pursued his work as a naturalist at the same time. One project was the acclimatization of rubber trees to Malaysia. He died in 1905, aged 80, in Italy.

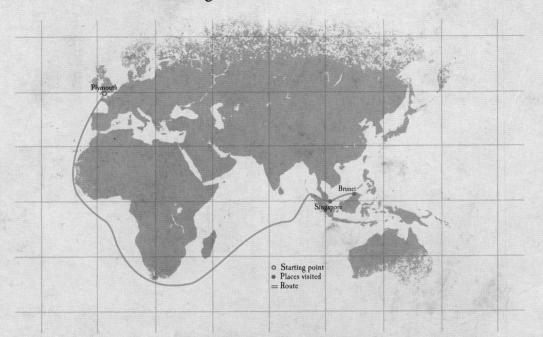

○ Starting point
● Places visited
═ Route

*Bamboo bridge above the village of Sennah, an engraving
from* Sarawak: its inhabitants and productions: being notes
during a residence in that country with H.H. the Rajah
Brooke, *by Sir Hugh Low, 1848*

The Malaysian version of *The Man Who Would Be King* began in 1839. James Brooke, former lieutenant in the British Indies, put all his money into a schooner, armed it, and went to seek his fortune in the Malay Archipelago. In Borneo, the sultan hired him as a mercenary to fight the Dayak rebels — known to be head-hunters — in Sarawak. The rebellion was soon put down and as a reward, James Brooke became the governor of Sarawak. In 1841 he proclaimed himself rajah, or king, and reigned as an enlightened dictator over a huge expanse of jungle. That is when Hugh Low arrived, a 20 year-old botanist with the East India Company. Low's botanical talents and his youth were both attractive to the homosexual White Rajah, and he became Brooke's private secretary and companion. For 30 months he collected seeds and plants in the interior of Sarawak where no European had ever ventured. Taking a great interest in the Dayaks, Low learned some of their dialects and made lists of the plant products they employed: rice, bamboo, wicker, cinnamon, pepper, camphor, areca (or betel) nuts and gutta-percha, the natural latex sap of the *Palaquium gutta* tree. Many had promising economic uses.

The primeval forests of Borneo held many inestimable species of trees and perfumed flowers, including jasmine, *Magnolia champaca* (champak), ylang-ylang, etc. The shores of the Sarawak River were astoundingly luxuriant. It was a paradise of orchids, and Low was able to introduce a number of new varieties to England. "The most gaudy are perhaps the various species of Cælogyne, called collectively by the natives the 'buñga kasih-an,' or the flowers of mercy; they are all highly fragrant, and their white and orange colored flowers are exceedingly delicate and beautiful." Perhaps because of the extreme humidity of that climate, rhododendrons become epiphytes and their roots "be-come large and fleshy, winding round the trunks of the forest trees; the most beautiful one is that which I have named in compliment to Mr. Brooke. Its large heads of flowers are produced in the greatest abundance throughout the year: they much exceed in size that of any known species, frequently being formed of eighteen flowers, which are of all shades, from pale and rich yellow to a rich reddish salmon color; in the sun, the flowers sparkle with a brilliancy resembling that of gold dust." But among his thousands of discoveries his greatest was the eight unknown species of carnivorous plants, *Nepenthes* or pitcher plants: enormous purple and scarlet pitchers with lids, climbing to the top of the trees, where they devoured insects and even birds.

CARNIVORE!

The largest carnivorous plant is *Nepenthes rajah*, discovered by Hugh Low on his first ascent of Mount Kinabalu, Borneo, in 1851. Its flower can be more than 12 inches (30 cm) long and contain 5 pints (2.5 l) of digestive juices. It is not unusual to find insects, frogs and even rats inside.

DYNASTY OF THE WHITE RAJAHS

The dynasty of the Rajahs Putih (white kings) lasted more than a century. In 1868 James Brooke, who had never married, passed the throne to his nephew Charles. When Charles died in 1917, his son Charles Vyner Brooke succeeded him, but he neglected Sarawak and it became a colony of the British Crown in 1946.

"On the banks of the rivers, and growing as underwood in the dense jungles, are found many beautiful species of the genera Ixora *and* Pavetta, *the former with large bunches of flowers of every shade, from orange to crimson, the latter with tufts of pure and delicate white blossoms ..."*

Hugh Low

FACING PAGE

HERBARIUM PLATE

Nepenthes rafflesiana

This specimen of a carnivorous Raffles Pitcher Plant was collected in Borneo by Low himself.

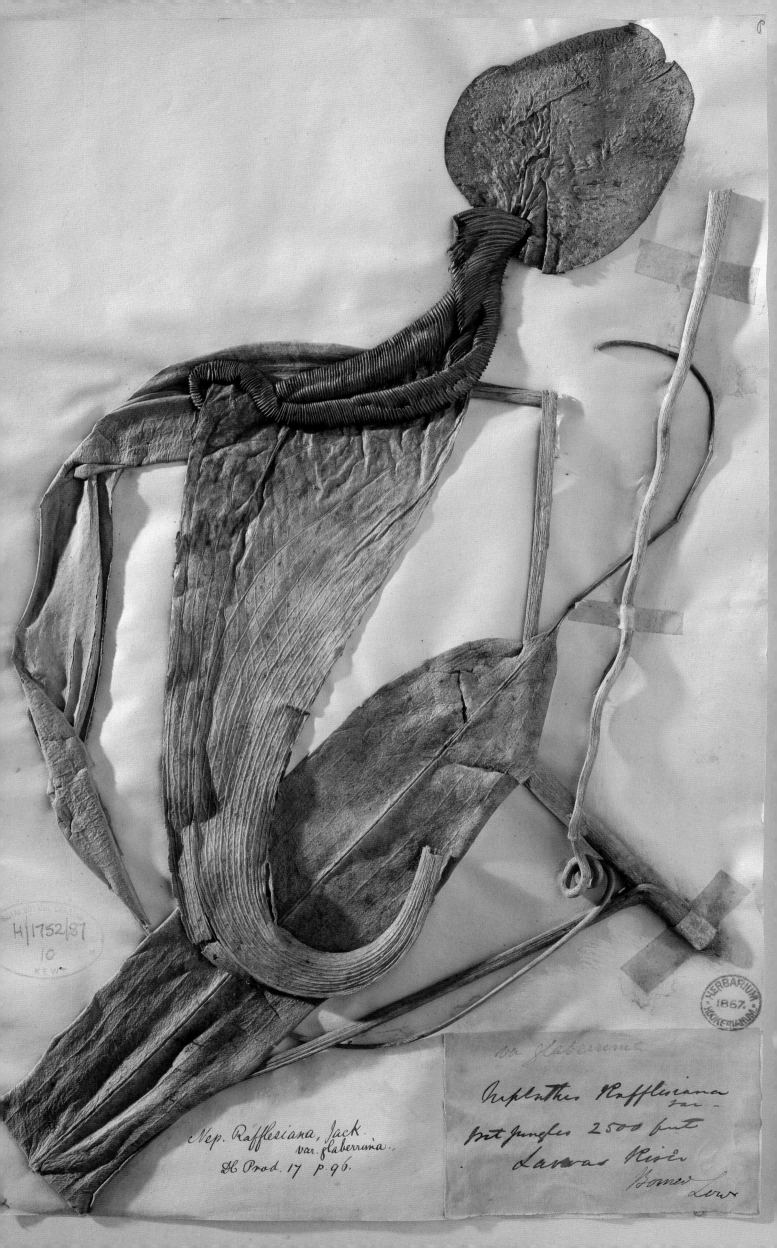

Nep. Rafflesiana, Jack.
var. glaberrima.
DC. Prod. 17 p.96.

var. glaberrima

Nepenthes Rafflesiana var -
Mt jungles 2500 feet
Lawas River
Borneo
Low

TRIBULATIONS OF A TEA THIEF

Robert Fortune in the Chinese Empire

Robert Fortune, born in Scotland, worked as a botanist at the Edinburgh botanical gardens before leaving on his first trip to China in 1842, at the age of 31. He was sent by the London Horticultural Society to find new plants. His second journey, in 1848, was an industrial espionage mission to discover how tea was grown, sponsored by the East India Company. He went to China twice more, in 1853 and 1858, and discovered how silkworms and rice were cultivated. He made a comfortable income for his retirement from publication of his travel journals, and died in London in 1880.

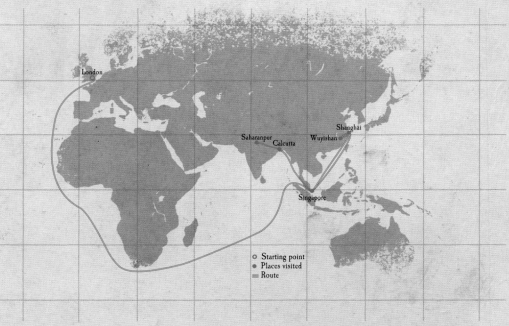

- ○ Starting point
- ● Places visited
- ═ Route

R obert Fortune's mission to bring back the best tea plants was undertaken in a tense political and economic time. The Middle Kingdom was willing to sell its tea to the tea-thirsty British, but jealously guarded the secret of growing the plant. Furthermore, payment had to be in silver. The British East India Company illegally flooded the Chinese market with opium in order to recover the silver bars. The Emperor resisted and tried to close his empire's borders. In the name of free trade, the British started the First Opium War.

In 1842, the Treaty of Nanking forced the humiliated Chinese to open five ports to foreigners. The finest tea plantations were still beyond the reach of Westerners, who were seen as "foreign devils." Robert Fortune ordered a "Chinese wardrobe," had his head shaved and wore a false pigtail down to his heels. On a previous voyage he had learned to speak Mandarin and how to handle chopsticks and considered himself a "very satisfactory Chinaman." With his servant, Wang, and a coolie, he went up the Yang-tseu-kiang (Yangtze) River on a trader's heavily-laden boat. His servants were quarrelsome and greedy, and nearly caused his unmasking and execution several times. Going up the many rapids of the Blue River took 12 days and Fortune gathered plants enthusiastically. When they arrived at the foot of the sacred mountain Sung-lo-shan, "the hill where green tea is said to have been first discovered" by a priest, he spent his days feverishly collecting seeds and shoots of the tea plant, observing how it was grown, and gathering information. On the return journey, he survived a fall in his chair (being carried by a porter) and a flood tide, but calmly carried on to Silver Island to continue his collecting. From Shanghai, he set out again, from canal to canal, and by

A tea plantation. Engraving from A Journey to the Tea Countries of China, *by Robert Fortune, 1852*

chair, and finally on foot, to the Bohea mountains, where the "best black tea" was grown. The monks there welcomed him and he learned the importance of water purity in preparing tea and also acquired 400 young plants. With their roots packed in damp moss, they were wrapped in oiled paper and sent to Shanghai. Fortune discovered how to ship seeds and plants in Wardian cases, those mini-greenhouses or terrariums, to remain healthy. In 1851, 12,838 tea plants were transplanted to the Himalayan foothills, via Calcutta, and Fortune had also recruited eight expert Chinese tea workers. Thus the first crack was opened in a 5,000-year monopoly.

GREEN TEA, BLACK TEA

In the 19th century, green and black tea were thought to be different species of plants. Robert Fortune discovered that they came from the same plant, *Camellia sinensis*, and that black tea is merely a fermented form of green tea.

FLOWERS AND TREES

Fortune provided Europe with access to many species of peonies, azaleas and chrysanthemums, and a yellow climbing rose (*Rosa fortunei*) stolen from the imperial city of Souchow, as well as trees such as the weeping cypress, windmill palm and kumquat.

"The long trains of coolies laden with chests of tea and other produce, and with the mountain chairs of travelers, presented a busy and curious scene, as they toiled up the mountain side, or were seen winding their way through the valleys."

Robert Fortune

FACING PAGE

HERBARIUM PLATE

Camellia sinensis formerly
Thea bohea

These specimens of tea were collected in China by Robert Fortune himself in 1845.

Thea Bohea

R Fortune 1605
—— Ahenos
—— 1845

TREASURES OF THE PRIMEVAL FOREST
Richard Spruce in the Amazon

Southampton

Casiquiare Canal

Ambato

Manaus   Belém

Tarapoto

○ Starting point
● Places visited
═ Route

Richard Spruce was born to a modest family in Yorkshire in 1817. He was taught by his father, a schoolmaster, and showed precocious talent for botany. At 16 he made a list of the plants in his village and at 19, he created a Flora for the surrounding district. Although he became a professor of mathematics, Spruce continued his botanical pursuits and specialized in bryophytes (mosses and sphagnum). His growing reputation made it possible to organize a collecting trip to the Pyrenees in 1845–46, and then to be chosen by Kew Gardens for an expedition to the Amazon. Despite his frail health, Spruce spent 15 years in South America, where he did exceptional research. He died in Yorkshire in 1893.

To see the tropical forests before it was too late! That was the burning desire of botanist Richard Spruce who, although suffering from tuberculosis, decided to go to Brazil on behalf of the Royal Gardens at Kew in 1849. He disembarked at the mouth of the greatest river in the world that runs through the vastest of forests and began collecting immediately. The conditions were horrifying. His hut was "stocked with rats, vampires, scorpions, cockroaches ..." and his specimens went moldy overnight. But Spruce carried on and retained his sense of humor. He spent a year exploring the forest around Manaus, 3 years going up the Rio Negro and studying its flora, then entered Venezuela. In 1854 he went up the Amazon on a steamer as far as Peru and continued in a canoe to Tarapoto at the foot of the Andes. After 2 years, Spruce's collection of plants was unimaginably rich. He had 250 different kinds of ferns, gathered within just 19 miles (30 km) of the village.

There he was in late 1857 when he received a message from Her Majesty Queen Victoria asking him to find specimens of *Cinchona pubescens*. This tree was the source of quinine, which was the only cure for malaria then ravaging the British colonies. Using all his diplomacy, Spruce was permitted to gather red cinchona in the forests of Ecuador. It took him 14 perilous weeks to get there by dugout canoe, over swollen rivers and in great hunger. There, although his back and legs were paralyzed, he did important work on the cinchona bark, collected 100,000 seeds and produced 637 plants. These precious shipments to England gave rise to immense plantations that saved thousands of lives. Spruce had a real gift for discovering rare species and also found several species of *Hevea brasiliensis*, the rubber tree, in the midst of the economic boom in rubber. His notebooks were full of observations on the plants with economic value: gums, resins, fibers, foods, nar-

cotics and stimulants, dyes and lumber. Spruce was rewarded with the royal sum of £100 a year in recognition of his successes.

In 1864, his failing health obliged him to return to England after 15 years away, but his heart remained in the Amazon. "I well recollect how the banks of the river had become clad with flowers, as it were by some sudden magic, and how I said to myself, as I scanned the lofty trees with wistful and disappointed eyes, 'There goes a new *Dipteryx* — there goes a new *Qualea* — there goes a new 'the Lord knows what!' until I could no longer bear the sight, and covering up my face with my hands, I resigned myself to the sorrowful reflection that I must leave all these fine things."

Gustavia pulchra painted by Margaret Mee

SHAMANIC PLANTS

Richard Spruce created deep and respectful relationships with some Amazon tribes. The Indians told him the secrets of plants such as *Banisteriopsis caapi*, a vine or liana rich in alkaloids, and *Anadenanthera peregrina*, a legume whose seeds, ground up and taken as snuff, contain tryptamines. These entheogenic substances (psychoactive chemicals for religious purposes) were used to produce shamanic visions of flying, death, rebirth or metamorphosis into animals.

FACING PAGE

HERBARIUM PLATE

Gustavia pulchra formerly
Gustavia speciosa
This specimen was collected on the Rio Negro in the Amazon region by Spruce himself in 1852.

"At Tauaú, I first realised my idea of a primeval forest. There were enormous trees, crowned with magnificent foliage, decked with fantastic parasites, and hung over with lianas, which varied in thickness from slender threads to huge python-like masses, were now round, now flattened, now knotted, and now twisted with the regularity of a cable."

Richard Spruce

LECTOTYPE

Gustavia pulchra Miers

ected by S. Mori

Flora Neotrop., Monogr. 21, Part I : 182
(1979)

Gustavia pulchra Miers

LECTOTYPE

Det. Scott Mori   197 3
Herbarium, University of Wisconsin

1933 Gustavia
Rio Negro — gapó abou Barcellos
Cal. very contrite acute 6-lob   Dec/51
so-sy small bushy tree.
Fl. rose very odong?
The fl. are infested by legions of caterpillars so
that it is scarcely possible to dry one entire.
Gustavia   Spruce 1852

Gustavia Speciosa

G. pulchra Miers
in Linnaea 1852

THE WAR OF THE ORCHIDS
Benedikt Roezl in Latin America

Benedikt Roezl was born in Bohemia in 1824. His father was a gardener and young Benedikt began his apprenticeship at 13 in the gardens of the Count of Thun in Bohemia. He continued his career in prestigious private gardens in Galicia, Moravia and Belgium, where he became head of cultivation at the national horticultural school. But he could not "restrain any longer his ardent desire to see the tropics," and moved to Mexico in 1854. There he created a nursery and grew and sold conifers. Then, for 20 years he crisscrossed the Americas hunting orchids for Sander & Co. After a life filled with adventures, Roezl retired to Prague, where he died in 1885.

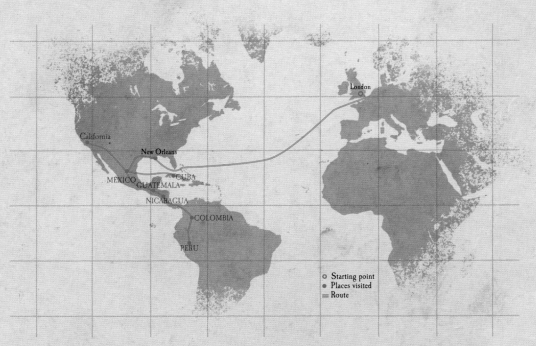

London

California

New Orleans

MEXICO   CUBA
GUATEMALA
NICARAGUA
COLOMBIA
PERU

○ Starting point
● Places visited
═ Route

In the 1870s, during Queen Victoria's long reign, Great Britain was seized by a passion for orchids. The gentry were buying plants at high prices and collectors were driven by the idea of profit. Theft, espionage, even murder — nothing was beyond possibility for those who wanted to own an unknown species or ruin a rival. The law of the jungle meant trees were felled just to get a specimen, and the whole colony of orchids destroyed to ensure the specimen was unique. A less radical, and perhaps more elegant solution was to urinate on a rival's bagged plants as they were being loaded on the ship, ensuring that they would be rotten when they arrived.

Benedikt Roezl was tall, with a full beard and an iron hook in place of his right hand; he was in no way a tenderfoot. He was the most intrepid and most pitiless of all the orchid hunters. Nothing stopped him; not "bands of thieves" nor "wild Indians." He was attacked 17 times, but his cunning and humor helped him escape. On behalf of his employer, Henry Sander, the keen founder of Sander & Co., Roezl searched, on foot and on horseback, through the forests of Central America and the west coast of North America. In 1869, he shipped more than 10,000 orchids to Europe from Peru and Colombia. From Caracas, he sent 9 tons (8 tonnes) of orchids and 11 tons (10 tonnes) of other plants to London. In Mexico, near the Colima volcano, he gave a reward to the Indians who brought him more than 100,000 plants. But the losses were enormous. Stuffed into baskets and crates, carried roughly on the backs of mules, the rare *Telepogon* specimens he collected at more than 11,000 feet (3,300 m) in the Andes all died before reaching the jungle. The

long sea voyage was often fatal to the most fragile species. Of the 27,000 *Dracula* or "vampire" orchids that left Colombia, only two plants arrived alive in England.

Like any good hunter, Roezl prided himself on his sixth sense. When he was in London, he paid close attention to the auctions, examining the new species on offer and guessing where they came from. Several months later, he hunted them down in the depths of the rainforest. But his greatest success was with the white *Cattleya*. He was convinced that these usually pink or mauve orchids would sometimes produce a root with a white flower. One day, he found one among the flowers that decorated a church in a distant village in Guatemala. He persuaded the padre and parishioners to exchange this *Flor de San Sebastian* for a secret that would ensure victory to the village's fighting cocks. Roezl eventually sold the rare orchid for the fine sum of 50 golden guineas.

ORCHIDMANIA

The fevered pursuit of orchids that gripped English society began in 1818, when collector William Swainson used dried *Cattleya labiata* plants as packing material for precious specimens from Brazil. The "stuffing" was covered in glorious pink and purple flowers when it arrived, and gardeners went mad.

THREATENED

The pressure of hunters and collectors on wild orchids decreased once gardeners mastered the ways to reproduce these plants. Still, many species have already become extinct or are under threat from deforestation and the disappearance of their pollinators: insects, bats and birds.

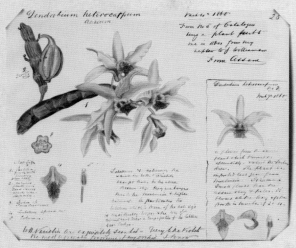

Drawing of Dendrobium heterocarpum *by John Day*

"In a very humid forest, through which we pass, all the trees were adorned with green moss, which pushed the adorable Odontoglossum cervantesii precisely in full bloom of astonishing diversity, from pink to pure white, and whose familiar design — gold, brown and purple — was extremely varied, as well as the size of the flowers."

Benedikt Roezl

FACING PAGE

HERBARIUM PLATE

Masdevallia roezlii or *Dracula roezlii*

These specimens of *Dracula* orchids were collected in Colombia in 1889.

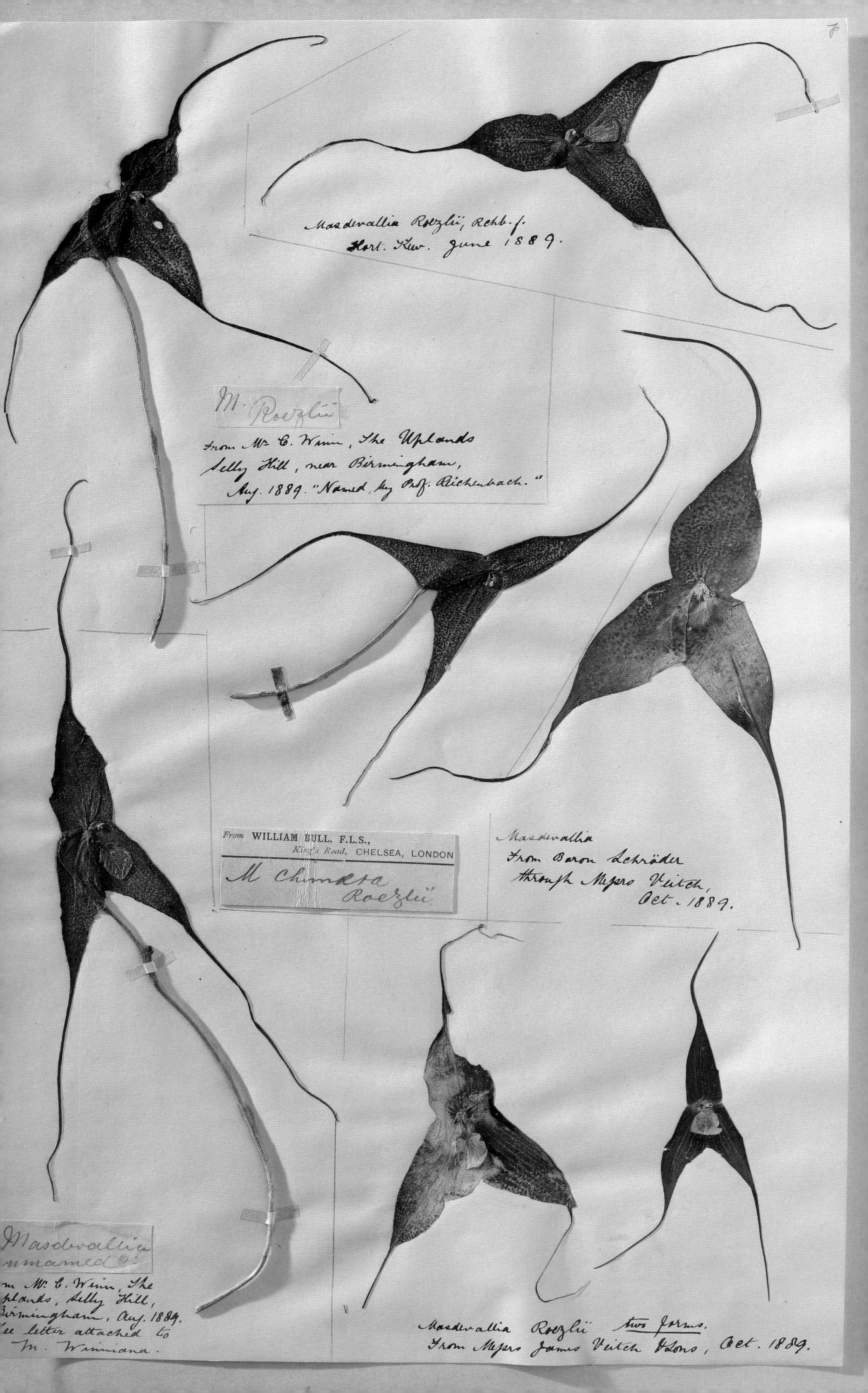

Masdevallia Roezlii, Rchb. f.
Hort. Kew. June 1889.

M. Roezlii

From Mr. E. Winn, The Uplands
Selly Hill, near Birmingham,
Aug. 1889. "Named by Prof. Reichenbach."

From WILLIAM BULL, F.L.S.,
King's Road, CHELSEA, LONDON

M. Chimaera
Roezlii

Masdevallia
From Baron Schröder
through Messrs Veitch,
Oct. 1889.

Masdevallia
unnamed

m Mr. E. Winn, The
plands, Selly Hill,
Birmingham, Aug. 1889.
ee letter attached to
M. Winniana.

Masdevallia Roezlii two forms.
From Messrs James Veitch & Sons, Oct. 1889.

Dr. BOTANIST, I PRESUME?
David Livingstone and John Kirk on the Zambezi

David Livingstone was born in Scotland in 1813; his family was so poor that he had to go to work in the cotton mill at the age of 10. But he studied at night and showed such intelligence that he was admitted to the University of Glasgow where he studied medicine and theology. He became a preacher with the London Missionary Society and was sent to Cape Town in 1841 and then to Botswana. From there, he succeeded in crossing Africa from west to east. After this success, Livingstone organized an expedition up the Zambezi River, from 1858 to 1864. He set out once more, in 1866, to find the source of the Nile, but illness forced him to remain on the shores of Lake Tanganyika, where Henry Stanley discovered him. He died in 1873 in Zambia but was buried at Westminster Abbey.

John Kirk was born in Scotland in 1832. His father was a pastor. John was a passionate student of plants and was elected to the Botanical Society of Edinburgh as a young man. Kirk earned a medical degree and practiced in Crimean War hospitals. While in the Dardanelles, he devoted his free time to botany and photography. In 1857 he was appointed surgeon and botanist to Livingstone's Zambezi expedition. His work was the basis of *Flora of Tropical Africa*, which appeared in 1868. That same year, Kirk moved to Zanzibar where he was the British Consul until 1886. His goals were to develop commerce and thus convince the Sultan to end the slave trade. He retired to England and died there in 1922.

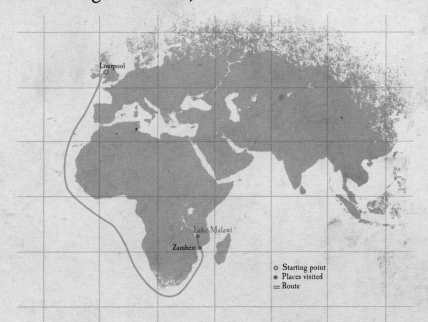

Liverpool

Lake Malawi

Zambezi

○ Starting point
● Places visited
— Route

Doctor David Livingstone was already a hero in 1858. As an Anglican missionary in central Africa, his conversions were few but his explorations were amazing. First, he crossed the terrifying Kalahari Desert, then he discovered spectacular waterfalls on the Zambezi River and named them in honor of Queen Victoria. Setting out from the Atlantic coast in Angola, he reached the Indian Ocean by descending the Zambezi River. Thus, he was the first European to cross Africa from west to east, and he accomplished that with a small group of indigenous helpers and despite having his left arm chewed by a lion. The Royal Geographical Society gave him its gold medal and supported his next project: he wanted to travel the entire course of the Zambezi by boat and prove that this immense river was the ideal route to penetrate the African continent and make use of its riches. With government support, the expedition acquired a steamer and took on several scientists. The botanist was a young doctor, John Kirk, recommended by the Director of the Royal Gardens at Kew, Joseph Hooker.

After leaving with great fanfare, the expedition soon ran into problems. Livingstone was a poor leader of men and an even worse organizer. The Portuguese were solidly in control of the delta mouths of the Zambezi and, to top it off, the river was not at all navigable above the Cahora Bassa rapids. They had to travel on the tributaries of the great river. Early in 1859, Livingstone and Kirk traveled along the Shire River and discovered the expanse of Lake Nyasa (Lake Malawi). It was a land of hostile tribes, slave hunters, savage beasts and malaria-carrying mosquitoes. Every evening at camp, three beds of dry grasses were laid out side by side; Doctor Livingstone was in the center, his brother Charles on his left, and Doctor Kirk on his right. "Our bags, our rifles and our re-

Portrait of Sultan Tipu Tipu, a famous warrior, who was also a merchant selling slaves and ivory. Photograph taken by John Kirk

volvers are carefully placed at our heads and a fire made near our feet," related the explorer. But that did not prevent the porters from robbing them and running away. Nevertheless, the botanist was successful in collecting plants that had not yet been described, including Zanzibar gem (*Zamioculcas*), *Cladostemon kirkii* and the African violet (*Saintpaulia*).

At Kebrasa Falls on the return journey, Kirk's canoe was carried away by a violent whirlpool and while the doctor escaped, clinging to a rocky outcrop, "All that was valuable, including a chronometer, a barometer, and, to our great sorrow, his notes of the

TOXIC TOOTHBRUSH

John Kirk accidentally discovered a powerful poison by slipping a cutting of *Strophanthus* into the pocket where he carried his toothbrush. After using the brush, and finding it rather bitter, he noticed that his pulse had slowed considerably.

SULTAN AND SLAVER

In the late 19th century, the fortunes of the Sultan of Zanzibar depended on trade in black slaves, ivory and spices. For years, John Kirk tried every argument to convince the sultan to stop selling humans. In 1873, when Kirk became Consul, he forced the sultan to sign a treaty, although it was frequently broken.

(continued on page 126)

August, 1864

"My dear Dr. Hooker,

I take the liberty of sending you a box of things … It contains the crania of two hippopotami, some bulbs and two small tusks — I think the bulbs, in particular one of the smaller species, will be of interest to you."

David Livingstone

FACING PAGE

HERBARIUM PLATE

Rhizophora mucronata

The specimen of red mangrove on the upper right was collected on the left bank of the Zambezi by Kirk himself in 1858, during his expedition with Livingstone. The drawings are also by Kirk.

Left Bank Luabo 46
15/5/58

Livingstons Exp.
Dr Kirk

Revisto para
"Flora de Moçambique"
por A.R. Gonçalves 1976.

Flora Zambesiaca

Rhizophora mucronata Lam.
(both specimens)
Det. W. L. Brealey 1974

Rhizophorae Juss.

Pd species Natalensis
rare!

Harv. & Sond Fl Cap. II 513

Prod. Natal. dedicit
1844 . W. Guinzius.

Rhizophora mucronata

The Lady Nyassa *being constructed, with the* Pioneer *at Shupanga, June 1862.*
Photograph by John Kirk

journey and botanical drawings of the fruit-trees of the interior, perished."

The two Europeans returned to the coast to receive a new paddlewheel steamer — in parts. The boat consisted of 24 heavy steel sections and it took 6 months to transport and assemble them. Said Livingstone, "Our progress was distressingly slow." The British government shared his opinion, after 3 years without results.

But going up the Ruvuma River was disastrous. The steamer's wheels got stuck in duckweed and churned up the bodies of slaves killed by slavers. Tragically, Livingstone's wife Mary died of fever despite Dr. Kirk's care, and Livingstone's temperament got worse. One by one, his assistants resigned or simply left, and John Kirk eventually wrote, "I can come to no other conclusion than that Dr. Livingstone is out of his mind ..."

In 1864, the government officially recalled the explorers. Kirk returned to England with 4,000 species of plants, specimens of the most precious wood, especially teak, various kinds of cotton, and too many insects and birds to count.

And yet, the two men were not finished with Africa. Kirk soon became British Consul in Zanzibar where he acclimatized a number of tropical spice plants, particularly *Cinchona officinalis*, the source of the antimalarial drug, quinine. As for Livingstone, he decided to seek the source of the Nile and left in 1866 with a small group of black servants. His oxen were wiped out by the tsetse fly and his porters fled, predicting his death. Alive but very ill, Livingstone made it as far as the Luapula River, which he thought was a tributary of the Nile. Skeletal, covered with ulcers, forced to beg

for food, Livingston arrived at Lake Tanganyika in 1871. That is where he was found by Henry Morton Stanley, a reporter with the *New York Herald Tribune*, who wrote, "As I advanced slowly towards him I noticed he was pale, that he looked wearied and wan, that he had grey whiskers and moustache, that he wore a bluish cloth cap with a faded gold band on a red ground round it, and that he had on a red-sleeved waistcoat, and a pair of grey tweed trousers. I would have run to him, only I was a coward in the presence of such a mob — would have embraced him, but that I did not know how he would receive me; so I did what moral cowardice and false pride suggested was the best thing — walked deliberately to him, took off my hat, and said:

'DR. LIVINGSTONE, I PRESUME?'

'Yes,' said he, with a kind, cordial smile, lifting his cap slightly."

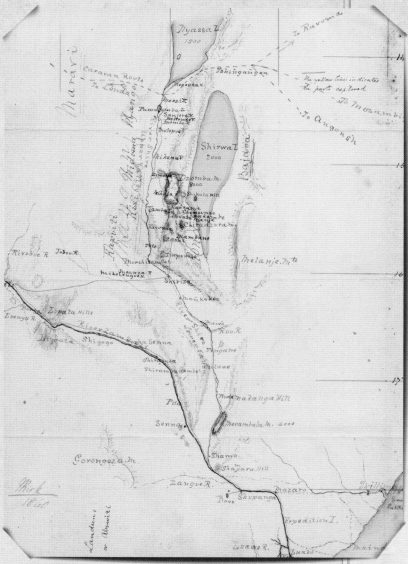

Map of Lake Nyasa, the lake of storms, drawn by John Kirk himself

"We were robbed in the night and ... now have only the clothes we were sleeping in. The thieves took all my specimens and, seeing they had no value, threw them away, but they were nearly destroyed by having been trodden into the sand."

John Kirk

Zig-3 39

Shire Valley opposite the
Rapids of Tetzane

Nov. 1861

Wk.

FLOWERS FROM THE "WHITE MAN'S GRAVE"

Gustav Mann in Cameroon

Gustav Mann, born in Hanover in 1836, became a gardener at Kew in 1859. Sent by Director William Hooker to replace a botanist who had fallen ill during an expedition in Niger, Mann eventually studied the vegetation in the Gulf of Guinea, Equatorial Guinea, and the islands of Principé, Saõ Tomé and Annobon. He and Richard Francis Burton were among the first Europeans to reach the summit of Mount Cameroon. Gustav Mann then joined the British Forest Service in India. He traveled to the tea plantations in Darjeeling and continued collecting, particularly ferns. He retired to Munich in 1891 and died in 1916.

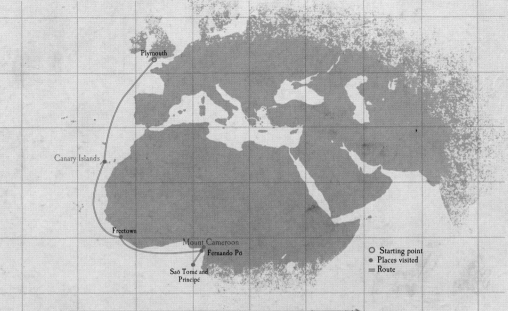

Plymouth

Canary Islands

Freetown

Mount Cameroon
Fernando Pó

Saõ Tomé and Principé

○ Starting point
● Places visited
═ Route

William Hooker, the director of Kew Gardens was very angry. He wrote to his friend Charles Darwin that "That impudent liar Burton has ... filched away all poor Mann's credit for the ascent of the Cameroons, calls it his Expedition, planned & carried out by him, & calls Mann his volunteer associate. I never read any thing so gross in my life — Poor Mann had set his heart on this thing for 2 years." It was impossible for Gustav Mann, a humble German gardener of just 25, to rival Richard Burton, the explorer who had found the source of the Nile. And yet, Mann showed his perseverance and his resistance to challenges. In 1860 Hooker sent him to join the Baikie expedition on the Niger River, replacing the botanist who had died of dysentery. Intestinal infections, yellow fever and malaria were taking such a toll on Europeans that West Africa was known as "the white man's grave." While waiting for news from Baikie, the gardener set out on a botanical exploration of the island of Fernando Pó and the coast of the Gulf of Guinea. In 1861 he attempted to climb Mount Cameroon but gave up because of the hostile tribal chiefs. He was in negotiations with them when Richard Burton, officially the British Consul, used his powers to get the expedition underway. The inexperienced Mann was passed by Burton on the first ascension to the peak of Mount Fako, 13,400 feet (4,090 m). Burton ridiculed the "*botaniker*'s" tin boxes, heavy with brown paper and wire, but used Mann's observations when telling the story of his travels.

The vegetation on this tropical volcano was exceptional. At the base, the forest was rich with dozens of species — bombax or kapok, raffia palm, plantain, oil palm, azobé (*Lophira alatra*), African teak, yellow mahogany and others. Its steep flanks were covered with dense bushes and flowering vines: "The ipomoea, a bright blue morning glory," a superb white clematis, a multitude of orchids. Mann found a geranium, various cucurbits including "bitter melon, with its red and yellow flowers and its little spiny fruit," "a perfumed flower very like honeysuckle, and the large yellow mallow known as hibiscus." They crossed vast fields of ferns and then, at the summit, Mann found buttercups, everlastings, broom and the delicate blue speedwell, *Veronica mannii*. As Mann collected, William Hooker was not neglecting his "poor" gardener and named some 349 botanical species and a number of genera for him.

Letter from Gustav Mann to Joseph Dalton Hooker, dated February 29, 1862. In this letter, Mann, who was on the Cameroon River and suffering from dysentery, lists the plants he has collected there.

APHRODISIAC PLUMS

In Cameroon, Gustav Mann discovered the *Pygeum africanum* (also *Prunus africana*), is a bitter chokecherry whose bark has been used by Africans for thousands of years as an aphrodisiac or a remedy for what they called "old man's disease." Today it is recognized as a treatment for prostate problems.

PALMS THAT CAUSE LAUGHTER

The natives of Mount Cameroon cut into the oil palm, *Elaeis guineensis*, every morning to drink its sap, and, noted Richard Burton, "during this operation they gave a strange laugh, a triumphant cry whose echo came back from the hills and resonated strangely through nature."

"Suddenly, the brush and forest disappeared, as if a lumberjack had felled them and, O! Joy! we emerged from the damp and sticky grasses that had imprisoned us. Our eyes were relieved to see a field of tender green, a dense mass of little mosses, and thick ferns of just one species, T. nephrolepis.*"*

Richard F. Burton & Gustav Mann

FACING PAGE

HERBARIUM PLATE

Renealmia mannii

This specimen was collected on the island of Fernando Pó in 1861 by Mann himself.

M. 483/92
6
KEW

FLORA OF TROPICAL AFRICA.

172

Renealmia
Fernando Po
Mann 1861

R. Mannii Hp. 4
Hook L. Pl. tab. 8. 1430.
vol. 15, p. 25
Type specimen!

Cited

All flowers in spirit

N 117 L.

Fernando Po 1861

HERBARIUM
1867

IN THE GREAT
LAKES SAVANNAHS

James Grant and John Speke in East Africa

James Augustus Grant was
born in 1827 in Nairn, Scotland.
The son of a pastor, he was
sent to college and then to the
University of Aberdeen, where he
studied chemistry, mathematics,
natural history and botany. He
joined the army in 1846 and
took part in conquering Punjab.
He became friends with John
Speke but Grant was wounded
and sent home to England in
1858. Learning of Speke's plan
to search for the sources of the
Nile, Grant offered to accompany
him. When they returned, Grant
received the gold medal of the
Royal Geographical Society and
published his journal from the
trip, *A Walk across Africa*. In
1868 he took part in the British
expedition to Abyssinia, then
retired from the army. He died in
his home town in 1892.

○ Starting point
● Places visited
= Route

"You have had a long walk, Captain Grant," said
Lord Palmerston, with British understatement.
Queen Victoria's prime minister was commenting on
the 1,146-day journey to the sources of the Nile undertaken by James Grant and John Hanning Speke. The
latter, in his previous expedition in 1858, had already
seen a "vast expanse of water" and named it Lake Victoria. He was convinced he had solved the puzzle of
the source of the Nile, which had been bothering Europeans for centuries. But Richard Burton, the explorer
he had traveled with, contested that discovery, which
he had not seen with his own eyes. Since then, the two
men were at daggers drawn. Speke decided to go again
in 1860, this time with his faithful friend from the Indian campaigns, Captain James Augustus Grant.

With his background in natural sciences, Grant
took on the role of the expedition's botanist and an-
thropologist. They left Zanzibar in September 1860
and headed northwest, with a caravan of 115 porters,
64 soldiers and 11 pack mules carrying the arms and
ammunition with which Grant and Speke could shoot
big game animals. Zebras,
giraffes, antelope, hippopotamus and black rhinoceros
all fell before these hunters.
When game was scarce, hunger slowed the caravan. In early 1861 the men were struck
by fever, delirium, headache,
blindness and other symp-
toms. For several months,
Speke went away to engage
new porters, then they resumed their walk to the north,
between Lake Tanganyika
and Lake Victoria. Grant

observed the vegetation: aloes, acacia, spiny mimosas
and wild coffee bushes on the savannah, papyrus in the
swamps and mosses around waterfalls. He found one
unknown type of acacia, and new, strange fruits, one
"the color and size of the Indian loquat," another "scarlet" and as "refreshing as a lime." But Grant was soon
immobilized for 6 months by a nasty abscess on his leg.
To distract him, children brought him "plants in flower, birds' nests, eggs or other things" and he sketched
them. Speke left alone and finally found the Ripon falls,
the exact spot where Lake Victoria becomes the White
Nile. The two men were not reunited until August 1862
and began the slow trip down the Nile, by boat and on
foot, arriving in Cairo in March 1863.

In three years, Grant had accumulated many
drawings, anthropological and zoological observations, weather measurements and botanical collections. Modestly, he turned most of his work over to
Speke, but Speke, just before debating Richard Burton in front of the Royal Geographical Society, died in
a hunting accident that remains mysterious.

LOCAL CHARMS

The African healers tried to heal
the ulcers on Grant's leg. They
applied a hot poultice of cow
dung, salt and mud. They tied
a small piece of lava around his
ankle and, that not working,
another charm of wood and
goat's flesh.

HEAVY LOSSES

Speke and Grant's expedition lost
six white soldiers (one killed, five
ill), 30 warriors out of 64, 113 of
the 115 porters (desertion), all
11 mules and 15 goats out of
20 (stolen).

"Speke introduces Grant to the Queen-dowager of Uganda." Engraving from the Journal
of the discovery of the source of the Nile *by John Hanning Speke, 1863*

*"Thorny shrubs, cactus, climbing aloes, with pink flowers, covered them, or the jungle
of grass was varied by circles of brushwood, giving shade to the rhinoceros; the older
trees were veiled over with silvery grey moss, which drooped gracefully, like the
pendent branches of the weeping willow."*

James Augustus Grant

FACING PAGE

HERBARIUM PLATE

Margaretta rosea

These specimens were collected
in 1862 by Captains Grant and
Speke themselves. The plant was
named for Grant's wife, Margaret.

FLORA OF TROPICAL AFRICA.

Margaretta rosea, Oliv. in Trans. Linn. Soc.

Expedition to the Sources of the Nile. 1860-1863.

531

Unyoro, Uganda
"Petals & coronal-segments pink"

Collected by Capt's Speke and Grant.

531
Found growing in a sesamum field. tap-root. Unyoro 29 July 62

HERBARIUM HOOKERIANUM 1867.

THE MISSIONARY WHO WAS DEVOTED TO BOTANY

Père David in China

Armand David was born in 1826 near Bayonne, and from his father, a doctor, he learned to love "all beasts, birds and flowers" and "rambling over hills and valleys." When he finished college, he wanted to join the priesthood and studied philosophy before becoming a novice with the Congregation of the Mission, missionaries of Saint Vincent de Paul. As an abbot, he taught natural science for 10 years in Italy, but dreamed of going to convert "the poor Chinese." Sent to China in 1862, he made three journeys of exploration as a naturalist and correspondent for the Muséum and the Academy of Science. Exhausted and ill, Armand David returned to France in 1874 and died in 1900.

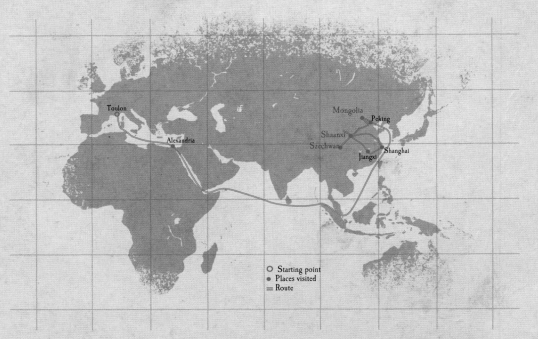

○ Starting point
● Places visited
═ Route

The First Opium War ended with the Treaty of Nanking, in which France and Britain forced a humiliated China to open up trade with the west. The victors demanded free circulation on the Yangtze River, access to five new ports for international trade and the right of Christian missionaries to spread their message to the people. The cathedral in Beijing (Peking) had been closed for 22 years and had just reopened when Father Armand David arrived in 1862. He was determined to "win over the immense population of the Far East for Christian civilization." He was given other missions by a number of scholars in the Academy of Science; they asked him to collect minerals, birds and animals, plants and seeds.

Père David set right to work: "While studying the language of the country and helping in our sacerdotal mission, I set out to explore the area around the capital." He went to see a Christian village in Mongolia, then a mountain massif near the summer residence of the emperor, north of Peking. His collecting was so productive and so carefully done that the Muséum authorities asked his superiors that he be allowed "for several years, to undertake grand voyages of exploration in the lesser known provinces of the vast empire."

On his first major expedition in 1866, he spent 8 months "in the middle of sad Mongolia," where "no European has ever set foot." Although what he collected was interesting, the missionary-naturalist was disappointed. "I have spent a lot of money and wasted my time and my labors, because the land is very poor, although people in Beijing told me otherwise," he complained in a letter.

He did not become discouraged, however, and prepared a second, longer expedition. In 1868 he went up the Yangtze (Blue) River on a variety of boats including a French gunboat. During the Taiping rebel-

Father Armand David in Chinese summer clothes, photographed in Beijing

lion it was a troubled time, and the roads infested with thieves. Christians were often harassed, and sometimes massacred. Père David finally found himself on a boat overloaded with "smoking tobacco, rice for winemaking, and even pottery" reeking of the cargo's odors and of the opium smoked by the boatmen "night and day." It was a difficult trip: the river was full of rapids, waterfalls and gorges. Dozens of haulers pulled the barge, attached to 1,300 foot (400 m) long ropes. The ropes would break, the boat was swept away and nearly over-

(continued on page 134)

DEER AND PANDA

With zoological expertise, Père David described 807 species of birds and discovered 60 mammals, including three antelopes, monkeys, rodents and especially Père David's Deer (*Elaphurus davidianus*), which was already near extinction in China at that time. The only survivors lived in an imperial park and he was able to get a skin by bribing the Tatar guards. In the foothills of the Himalayas, he also found the giant panda, a white bear "except for the limbs, the ears, and rings around the eyes, of a deep black, which lives on roots and bamboo."

"It is sad to see the speed with which these primeval forests are destroyed. There are nothing but tatters left of them all over China, and they will never be replaced."

Père Armand David

═══ FACING PAGE ═══

HERBARIUM PLATE

Pedicularis davdii

This specimen was collected in eastern Tibet, in the province of Moupin, by Armand David himself in 1870.

The life of missionaries in China in the 1850s was not without danger.
Torture and execution justified the military intervention by European countries.

MARTYRS FOR BOTANY

Many French missionaries devoted themselves to Chinese botany at great risk. Father Jean Marie Delavay went on foot through the mountains and sent over 200,000 specimens to the Muséum, even though one of his arms became paralyzed in 1888. Another was Father Jean André Soulié who disguised himself as a Chinese merchant to explore China's southwest from 1880 to 1890. He collected over 7,000 species, including a white rose from Tibet, *Rosa soulieana*, and the first seeds of *Buddleia*, butterfly bush. In 1905 Soulié was captured by Tibetan monks, tortured for 10 days and shot. His assistant, Father Bourdonnec, was decapitated.

turned. Père David decided to go by sedan chair to visit a little college of the Mission in Tibet, in the principality of Moupin. He went on many excursions from there, but in spite of his solid constitution and his "iron legs of a real Basque," he suffered from fevers and unbearable leg pains. His herbarium grew by over 150 new species, including the *Davidia involucrata,* the dove tree or handkerchief tree, with its spectacular white bracts, and the *Buddleia davidii,* the butterfly bush, as well as cotoneaster, primrose, strawberry, willow, violet, lilies and at least a dozen rhododendrons.

In 1870 David returned to France to regain his health, but landed in the middle of the Prussian occupation. When he arrived in Paris, he was pleased that "neither the 80 Prussian shells that had fallen in the Jardin des Plantes, nor the petrol of the Communards" had harmed the Muséum or his precious shipments. He returned to Shanghai in January 1872, ready for his third and final journey to the center of China. This time, he spent 32 days shaken in a terrible wagon, when he wasn't walking out of compassion for the pack mules floundering in the mud. Dust storms, floods, bandits, fevers, tigers and panthers — he saw it all. Even at rest stops there was little rest. In the inns, it was the height of "luxury if the roof was not open to the wind, if there were not the usual three kinds of vermin, if the nearby outhouses were not too smelly, and if the breath of the opium smokers was not too close."

In the end, he became ill from some "kind of plague or forest fever." He struggled heroically, said mass every day, was pulled by a mule on a 6 hour walk, and packed all his discoveries for shipment to the Muséum. They were numerous: insects, fish, birds, reptiles and mammals, alive or stuffed or preserved in alcohol, and hundreds of seeds and plants. He set out with his precious crates on the Yellow River, "a very nasty river" 6 miles (10 km) wide. He

was captured by pirates. The man of God had to fire his revolver and threaten to shoot the pot where the rice was boiling before he could get away. Further downstream, his boat foundered on a rock and he lost more than 1,000 plants and his giant salamanders (alive, in a crate). Harassed, underfed, he succumbed to fever, and in his bed he received the last rites, but recovered enough energy to get to Shanghai. In 1874 Père David left China for the last time, full of "regret to leave his explorations unfinished: Man proposes, God disposes."

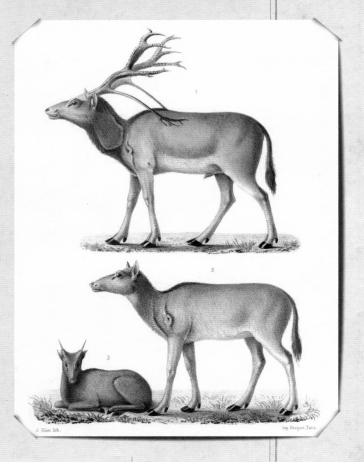

"*As for food, I depend on the Chinese, and I believe that with a little goodwill, one man can live wherever another can. I do not burden myself, therefore, with any food.*"

Père Armand David

FACING PAGE

HERBARIUM PLATE

Astilbe chinensis max. i davidii

These specimens were collected near streams in Gehol, China, by Armand David himself in 1864.

Mongolie orient. – Gehol
Juillet, 1914
Pr. Haute A. David
(?) De Richeaux fontaine

162f

1931

HERB. MUS. PARIS.

Castilla chinensis Max.
r. Davidi. Franch.

Gehol, près ravines montagne, fl. Juin 1864
CHINE (MONGOLIE ORIENTALE)
M. l'abbé A. DAVID. — Déc? 13/82.

FLOWER IN A GUN BARREL
Clovis Thorel in Indochina

Clovis Thorel was born in the Somme region in 1833, into a modest family of textile workers. His father, Charlemagne Thorel expected him to work in trade, but at 17 Clovis met a medical student in the Amiens botanical garden and that began his vocation. Against his parent's advice, he worked hard to study medicine, paying for his studies by working as a laboratory technician. In 1861 he joined the navy as a surgeon and was sent to Cochinchina. He patrolled the Mekong River and worked at the hospital in Saigon. After an expedition to the source of the Mekong, Clovis Thorel opened a medical practice in Paris, and then studied the waters of the hot springs at Bagnoles-de-l'Orne until his death in 1911.

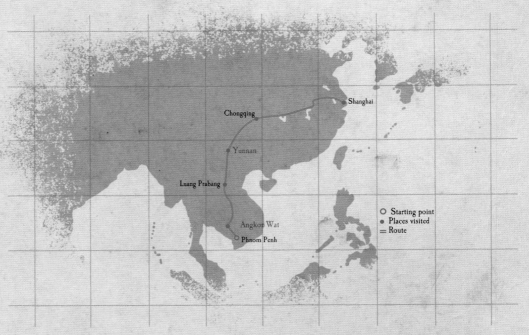

○ Starting point
● Places visited
═ Route

"At noon on July 7, all our preparations were complete; Gunboat 27, carrying all the personnel and equipment for the expedition, and Gunboat 32 set out together from Phnom Penh." To shouts of "Vive l'empereur! Vive le commandant de Lagrée!" the two military vessels began the slow journey up the Mekong River. The expedition's purposes included geography and natural sciences, but the prime mission was colonization. In the same year, 1866, France had defended its protectorate of Cochinchina (the Mekong delta) at gunpoint and intended to extend a "civilizing influence" over Siam, Laos, Burma and even China. Captain Doudart de Lagrée's mission was to follow the course of the Mekong to open a route to the Middle Kingdom that did not go through Beijing.

On board Gunboat 27 were the six members of the Mekong Exploration Commission, including Clovis Thorel, age 31, naval surgeon, botanist and a "tireless explorer of forests and arroyos." After a stop at the temple of Angkor Wat, hidden in vegetation, the expedition continued up the increasingly tumultuous river. Huge waterfalls meant the gunboats had to be left behind and the expedition continued in dugout canoes that had to be hauled through the rapids. Once they reached Cambodia the rains were torrential and the explorers were plagued by fevers and dysentery. Having "used up our entire stock of shoes," they were soon walking barefoot on stinging grasses, sharp rocks, leech-infested forests and through the hunting grounds of leopards and "Lord tiger."

Above Luang Prabang in Laos, the Mekong runs between steep cliffs of clay and marble, so the expedition left the river and climbed the high plateaus of the Chinese borderlands, crossed the Tonkin River and arrived at Yunnan. During this epic journey, Clovis Thorel made nine herbaria holding 4,200 species of orchids, poisonous plants and essences of tropical trees. He noted the therapeutic properties of each, as reported by the natives, with a precision that made him one of the pioneers of ethnopharmacology. As part of the colonization mission, he also made notes on resources such as spices, food plants, textiles and dyestuffs.

Despite the death of Captain de Lagrée, the sick and exhausted men pressed on to the Blue River and the Tibetan border, and reached Shanghai in June 1868. In 2 years they had covered 5,500 miles (8,800 km) and opened the mythical southern route from the French protectorates to China.

HOLY BOOKS MADE OF PALM LEAVES

In the ruins of the royal palace at Vientiane, the explorers climbed to the worm-eaten columns of the temple library and discovered sacred books lying on the ground. These were made of "long, narrow strips, cut from the leaves of a particular species of palm, gilded and gathered into books." Each of the Europeans took one specimen and "carefully hid it at the bottom of his bag, to hide this theft the natives would have considered a sacrilege."

In 1866 the mission led by Doudart de Lagrée stopped at Angkor, where de Lagrée took inventory of the major buildings and sketched the first outlines of the temple.

"Three highly poisonous plants grow in the forests: two Strychnos are very abundant and their seeds are harvested in great quantities in Cambodia and sent to Saigon, then on to China where they have many therapeutic uses."

Clovis Thorel

FACING PAGE

HERBARIUM PLATE

Glycosmis citrifolia Willd
formerly *Glycosmis cochincinensis* Pierre

This specimen was collected in Cochinchina by Thorel himself between 1862 and 1866.

Rutaceae Aurantieae

Glycosmis cochinchinensis Oure.

COCHINCHINE. *Nha-met*

M. le Dr THOREL, 1862-1866.

FLOWERS OF PASSION
Marianne North, a lady, a painter and a globe trotter

Places visited
Route

Marianne North was born in Hastings in 1830. Her father was a member of Parliament and she received a proper education for a girl of her status: music, painting and trips to Europe. Her mother died young and made Marianne promise never to abandon her father. Thus, Miss North determined not to marry and to study nature painting. When her father died, she was 40 and decided to fulfill her dream of going "to some tropical country to paint its peculiar vegetation." For 14 years, from 1871 to 1885, she visited every continent at least once, from Ceylon to California, Japan to Cape Town, painting hundreds of plants. In 1886, Marianne North retired to a small house in Gloucestershire and died at 59 in 1890.

In 1871 an unusual lady disembarked from a steamer in Kingston, Jamaica. She had not worn crinolines when traveling for some years, and in her bags were an easel, blank canvases, a trunk full of notebooks, pens, brushes, inks and oils. At 40, having spent 14 years looking after her father, she was beginning a new life. She rented a house in Kingston and spent 5 months painting the exotic flowers so admired by visitors to Kew Gardens. "The mango-trees were just then covered with pink and yellow flowers, and the daturas, with their long white bells, bordered every stream. [She] was in a state of ecstasy, and hardly knew what to paint first." She painted until sundown, surrounded by the *Ipomœa bona nox*, passion flower, giant ferns and orchids. Hers was a rapid technique: after making a pen and ink sketch, she applied paint straight from the tube. Her descriptions reveal the eye of the artist and the rich depths of her palette. "The sago-palms were just then in full flower, with great bunches of pinkish coral branches coming out of the center of their crowns. The fruit when ripe is like green satin balls quilted with red silk." Observing the Australian hakea plant she noted that, "every spike of leaves was gradually shaded downwards — the leaves at the top salmon pink, those next yellow and orange, and so into brightest green, blue-green, and purplish gray." She painted, "crimson seeds," pearl-colored" buds, graceful trees "with scarlet bark," and mosses "tinted with metallic blue or copper color."

After North America and Jamaica, Marianne North traveled to Brazil, Tenerife, Japan, Singapore, Malaysia, Sri Lanka, India, Australia, New Zealand, South Africa, the Seychelles and, finally, Chile. Because of her father's connections she was welcomed everywhere, but preferred to escape polite society. She describes her life: "I got out of my window, only

a yard above the ground, and went down to the stable: all asleep too, and the sun rising so gloriously! I could not waste time, so took my painting things and walked off to finish my sketch." Intrepid, she traveled through forest ("a perfect world of wonders"), mountains and savannah on foot, on horseback, in wagons and canoes, climbing, wading and falling. Her brave explorations led her to discover unknown species. In Sarawak she discovered an enormous, carnivorous, bright pink pitcher plant, and in the Seychelles, a new tree that would be named after her, *Northea sechellana*. On her final trip to Chile the audacious gentlewoman risked being eaten by pumas to paint the *araucaria* or monkey-puzzle tree, and the great blue puya, its "noble flowers standing out like ghosts."

Pitcher plant, Nepenthes northiana, *painted by Marianne North*

GUEST OF THE WHITE RAJAH

In Sarawak, Marianne North was the guest of Rajah Charles Brooke and his wife. As she enjoyed the luxuriant Malaysian forest, she allowed herself the pleasure of painting in a canoe, "stretched out on [her] back and looking up at the tangled branches above."

A GENEROUS DONOR

At 50, Marianne North not only gave all her paintings to Kew, but also paid to have a pavilion built to house them. She decorated the North Gallery herself and installed her 832 works, which remain in the same places today.

FACING PAGE

HERBARIUM PLATE

Nepenthes decurreus Macfarlane or *Nepenthes macfarlanei*

This specimen of the carnivorous pitcher plant was collected by J. Hewitt in 1908.

Small imperfectly matured
capsule with like small
seeds.

ISOTYPE N. decurrens Macf.

DET M. Jebb   8 19 95

+ note

seeds

TYPE

THE RUBBER RUSH
Henry Wickham in the Brazilian forest

Henry Alexander Wickham was born in 1846 in London, where his father was a lawyer. At 20, he left for Nicaragua and collected exotic feathers for the fashion trade. He went back and forth between London, where he married his publisher's daughter, and the Amazon, where he collected seeds of the rubber tree for Kew Gardens in 1876. This was a rare success, for his other projects in coffee, tobacco and copra did not make him rich, and his failures stretched from Brazil to Honduras and Australia to New Guinea. He went bankrupt several times and lived in poverty for years. After 1911 he was finally rewarded for his contribution to the prospering rubber industry and knighted as well, but died in 1928.

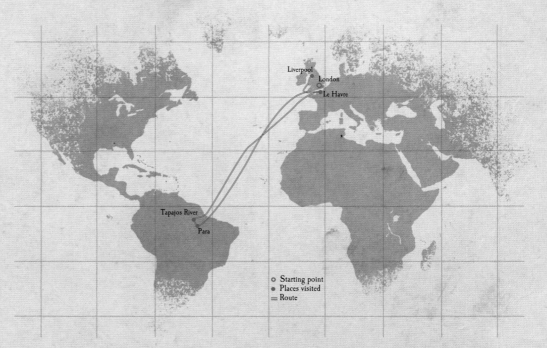

○ Starting point
● Places visited
═ Route

The world was hungry for rubber when Henry Wickham went out to seek his fortune. In 1839, Charles Goodyear had discovered that by adding sulfur and lead to the sap (latex) of the rubber tree, then heating it, a new material was produced that was fabulously malleable and stretchy. Soon, rubber was everywhere and vital to emerging industries such as tires, weatherproof garments, electrical insulators, weapons and more. Stock prices rose at a dizzying rate. Empires arose, based on enslaving the indigenous people of South America.

Henry Wickham had so far been a mediocre farmer, an unwise entrepreneur and an unlucky pioneer. His wife, Violet, accompanied him with resignation, through financial disaster and family losses. They were living at Santarem, Brazil, when their luck turned; In 1876, the British government asked Wickham to collect the seeds of *Hevea brasiliensis* and would pay £10 per 1,000 seeds. Wickham jumped at the chance. Learning that the steamer *Amazonas* had room in its hold, he immediately paddled up the Tapajos River in his canoe. He quickly hired some Indians and sent them out to scour the forest, where rubber trees grew wild and distant from each other. He had the village women weave baskets and line them with banana leaves for packing the dried seeds. Thus, 70,000 seeds were packaged in record time, loaded into dugout canoes, then onto the *Amazonas*, right under the noses of Brazilian authorities. Wickham boasted of his adventures and called himself "the thief at the end of the world," but while Brazil carefully maintained its monopoly on the production of latex, it had no laws or law enforcement to prevent exports.

Drawing by Wickham showing 30-year-old Hevea brasiliensis *in the Henarathgoda Botanical Gardens, Ceylon*

Of the 70,000 seeds sent to Kew, 2,397 germinated and were potted. More than 1,900 were sent to Ceylon, and others to Singapore, Jamaica, Australia and Cameroon. In 1892, one of the best specimens already had a circumference of 6.5 feet (1.95 m). Soon the dense, well-organized plantations in Asia had surpassed those in the Americas. When Wickham died in 1928, there were 80 million rubber trees growing in the British Empire.

Thus, Britain gained the first global monopoly on a raw material in high demand, but Wickham was paid only £700 for his seeds. Not long before he was about to die in abject poverty, he finally received £20,000 more for his "services to the rubber industry."

SLAVES TO RUBBER

In the latter part of the 19th century, one rubber plantation slave was dying for every automobile tire being produced. Murder, rape and deliberate famine were so prevalent that they led to the birth of the first human rights movement in the Congo in 1904.

FROM RAINCOATS TO TIRES

In 1823 Charles Macintosh of Scotland created the first rubberized raincoat. In 1839, Charles Goodyear of the United States invented vulcanization of rubber. In 1888, another Scot, John Boyd Dunlop, made the first pneumatic bicycle tire, and 6 years later, French brothers Édouard and André Michelin produced the first inflated automobile tire.

"*Santarem, Province of Para, Brazil, November 8, 1873*

Dr. Hooker, Royal Gardens, Kew

Dear Sir, I have just received a letter from Her Majesty's Consul in Para, inquiring what price I would ask to supply the government with seeds (in 50 kg lots) of the rubber tree to be introduced into India."

Henry Wickham

Siphonia brasiliensis Mill.

In vicinibus Para.
coll. R. Spruce,   Jul.-Aug. 1849.

Hvea brasiliensis (Willd ex Adr.
                    Juss.) Muell-Arg.

Determinavit *Schultes*   Aug. 1950.

THE WORLD WAS HIS GARDEN

David Fairchild, unstoppable agricultural explorer

David Fairchild was born in Michigan in 1869 to an academic family. After studying agronomy he joined the U.S. Department of Agriculture and worked on plant diseases. At 22, he helped create the Office of Seed and Plant Introduction, which he managed from 1904 to 1928. In 1898 he created an acclimatization garden and later a horticulture research station in Miami. He traveled around the world for 40 years and introduced thousands of foreign plants through these gardens. His books about his adventures, including *Exploring for Plants* and *The World Was My Garden,* were very popular. He died in 1945.

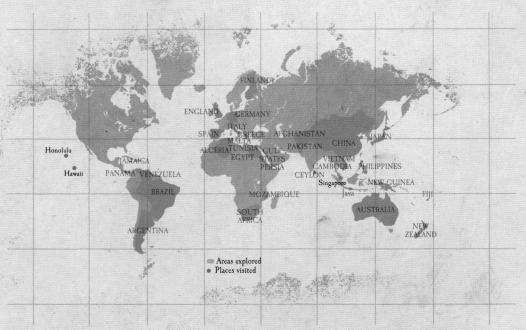

Areas explored
Places visited

A chance meeting between two Americans on an ocean voyage between New York and Naples in 1890 was the beginning of the greatest conquest of plants in the history of humanity. The wealthy traveler, Barbour Lathrop, became friends with David Fairchild, a talented young agronomist specializing in plant diseases. He persuaded the young man to travel with him and "see the world." Fairchild followed his demanding sponsor for four whirlwind and sometimes irritating years: a few days in Bali, a stop in Honolulu, a rest in San Francisco, then off again to Jamaica, Venezuela and on and on. Along the way, Fairchild made connections with naturalists and botanical gardens all around the globe.

"When the clock struck twelve and the new year of 1897 began, I had promised Mr. Lathrop that I would take up a study of the plants useful to man and, together with him, find a way to introduce their culture into America." Then they were off on 4 more years of relentless travel. This time, Fairchild had convinced the Agriculture Department to put him in charge of the Office of Seed and Plant Introduction, which at that time consisted simply of some land in Coconut Grove, just outside Miami, in Florida's mild climate. Fairchild sent his first discoveries there: from Jamaica, he sent chayote squash, from Venice, Sultana grapes with a promising future in California, from China the lychee fruit, from Egypt sesame and cotton that would flourish in Arizona, and from New Guinea the cajeput with its oil-rich leaves.

Until the 1930s, Fairchild dashed through Finland, Dalmatia, the Maghreb, Malta, Spain, England, China and Egypt again, around the Persian Gulf, then Africa, the Philippines, Vietnam, Japan, Afghanistan, Ceylon.... He found the time to fall in love with Marian Bell, the daughter of the inventor of the telephone, and married her in 1904. They settled in Florida, near the establishment where Fairchild's 100,000 plants were growing: rice, wheat, palms, bamboo, teak, long-fiber cotton, oil-producing plants such as palm and soy beans, and fruit trees of a dazzling variety: papaya, mango, avocado, sapodilla, orange, lemon, chestnut, peach, cherry, grape, pistachio, walnut and more. The prosperity of 20th-century American agriculture was rooted in Fairchild's gardens.

Fairchild examining a papaya tree in the experimental garden

PACKAGING

In the 1890s, Java was "almost as distant as the moon" from the United States. For his first shipments, Fairchild coated Javanese mangosteen seeds in paraffin and wrapped them in dry charcoal, a process that was successful at keeping the seeds alive.

CHERRY BLOSSOM FESTIVAL

In 1905, David Fairchild was acclimatizing some Japanese cherry trees in his gardens. In 1912 he coordinated the reception of a gift from the mayor of Tokyo to the city of Washington, D.C.: 3,000 Japanese cherry trees that were planted along the banks of the Potomac. Every spring since then, their dazzling pink blooms have attracted thousands of visitors for the National Cherry Blossom Festival.

"During my first expedition with Mr Lathrop, I sent back four varieties of Barbary fig or cactus (Opuntia): one came from Gran Chaco, Argentina, another from the chalky hills of Sicily, the third from Ceylon and the fourth from some very old stands on the Tunisian coast."

David Fairchild

FACING PAGE

HERBARIUM PLATE

Prunus yedoensis matsum

This specimen of cherry tree was collected in Japan in 1914 by E.H. Wilson.

No. 6362 ARNOLD ARBORETUM.

EXPEDITION TO JAPAN. 1914.

Hondo.

Prunus yedoensis

Matsum.

Alt.

Coll. E. H. Wilson. *April 2, 1914*

FLOWERS OF THE GOLD RUSH
The Lobel Family in the Klondike

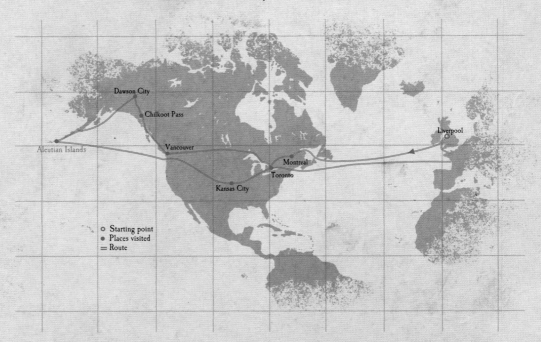

Posterity has little to say about Loicq de Lobel, and even less about his wife, who created the first herbarium of the Klondike. They were born in France near the end of the 19th century, and they had two boys and two girls. In 1898, pulled by the Gold Rush, the Lobel family made the long journey across North America. First they lived in Yukon, then they crossed Alaska and traveled as far as the Aleutian Islands. When they returned, Lobel had some success publishing the story of their travels. Then in 1904, styling himself Baron de Lobel, he represented some American magnates who convinced Czar Nicolas II to build a bridge across the Bering Strait. The Russian Revolution put an end to the project and Lobel faded into anonymity.

Dawson City
Chilkoot Pass
Liverpool
Vancouver
Aleutian Islands
Montreal
Toronto
Kansas City

○ Starting point
● Places visited
= Route

"Gold! There's gold in the Klondike! At the end of 1896 the news was spreading around the world and attracting hordes of international fortune hunters to the north. Among them was a young French entrepreneur, Loicq de Lobel, who hoped to sell prospectors a machine to concentrate the gold scattered in riverbeds. In April 1898 Lobel, his wife and four children joined the Gold Rush and headed for Dawson, "nicknamed the white elephant because it was an expensive, foolhardy endeavor to get there."

Like most gold seekers, the Lobels arrived in Alaska by boat and then had to go overland into Canada's Yukon Territory. They crossed swamps full of dead horses and arrived at the foot of the infamous Chilkoot Pass, a steep slope with 1,500 steps chiseled out of the ice. Miners went up the Pass single file, in a long, hungry, frozen line. When the Lobels arrived at the top, they found a lake where the ice was rapidly melting and dropped "with thanks to God" into the melted snow. They continued in a "frail skiff" called the *Lobelia*, which just managed to bear the weight of the family, baggage, three sled dogs and two hired men. Through storms, canyons and deadly rapids they continued, but the hired men abandoned them. "My daughters rowed the boat," wrote Lobel. Not only did the women endure these hardships with "extraordinary courage," Mrs. de Lobel managed to collect plants in this wild country. In July, they reached Lake Laberge; "We found red gooseberries, currants, delicious wild onions and anemones" but "for 40 km the water bubbles and foams in huge waves like those of the ocean."

In her herbarium, Mrs. de Lobel pressed the flowers she collected along the way: two kinds of lady's slipper orchid endemic to Alaska, astragalus bearberry, pink willowherb, arnica and northern bluebell. On August 7, they finally arrived in Dawson City, where the Yukon and Klondike rivers meet. In two years, a small fishing camp had grown to a city of over 30,000 inhabitants, with two banks and "2 fine barracks for the mounted police." Lobel admired the "most wonderful pure gold found in all of Alaska ... as big as a fist," and Mrs. de Lobel could finally enjoy the beauty of the landscape. "The mountains are full of brilliant flowers, superb wild roses, larkspur, forget-me-nots, lupins, sage, and mosses of all colors."

THE CHILKOOT PASS

At an altitude of 5,568 feet (1,697 m), this pass was a nightmare for the gold prospectors. Jack London made it famous in *The Call of the Wild*. Early in the Gold Rush, each prospector had to carry his own food, equipment and even dogs on his back. When the Lobels arrived, an aerial tramway had been installed to transport goods.

YUKON FORESTS

Lobel described the mountains of Alaska as being "very high and well wooded." He reported seeing fir, black pine, birch, poplar, laurel and ash, and several times mentioned a mysterious "enormous cotton tree."

A dog team in the Klondike, photograph from the newsmagazine Le Monde illustré, *March 9, 1901.*

"We resumed our march forward with difficulty. The snow became thinner, the vegetation more advanced and pretty flowers appeared everywhere ... arriving at Glenora, we found many blooming flowers, butterflies and birds. "

Loicq de Lobel

FACING PAGE

HERBARIUM PLATE

Cypripedium passerinum

This specimen of a lady's slipper orchid was collected by Richardson in Canada.

C. passerinum Rich
L. Richardson. Hook fl. Bor.
am. 2. tab. 206

IN PURSUIT OF THE REGAL LILY

"Chinese" Wilson in Szechuan

Ernest Henry Wilson was born in England in 1876. He apprenticed with a nursery and then worked in the Birmingham Botanical Gardens while taking evening courses. He won the Queen's Prize for botany. In 1897 he began work at the Royal Botanic Gardens, Kew, where he won the Hooker Prize for an essay on conifers. Wilson then accepted a position as Chinese plant collector with Veitch nurseries, and later for the Arnold Arboretum at Harvard University, Boston. From 1900 to 1927 he crisscrossed China, Japan, Korea, Formosa, then Australia, New Zealand, India, South America and East Africa. In 1927 he became head of the Arnold Arboretum and in 1930 he died in an automobile accident.

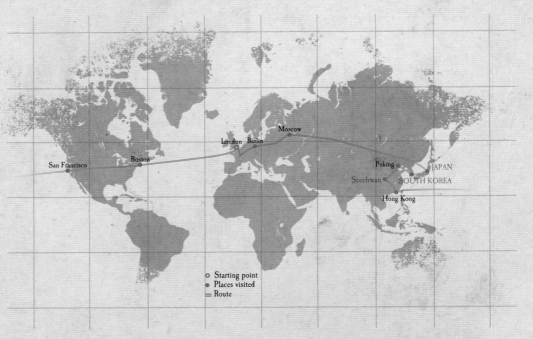

○ Starting point
● Places visited
▬ Route

In 1900, Ernest Wilson, a young employee of the Veitch nurseries in England, landed in Hong Kong on a specific mission: he was to find the near-mythical "dove tree," the spectacular *Davidia involucrata*, discovered by Père David 30 years earlier. All that Europeans had seen of it was a few dried flowers. The most recent had been sent by Veitch's Chinese correspondent, Dr. Augustine Henry, in 1891. When Wilson arrived, the only clue Dr. Henry gave him was a hand-drawn map showing an area of 19,300 square miles (50,000 sq km) marked with an "X."

Full of ambition, Wilson set up an expedition, plunged into the forests, found the exact valley, and 10 months later, found a fresh stump and a brand-new cabin made with the precious wood of the dove tree. His could not sleep at night until he finally found his holy grail, in full bloom, with its white flowers blowing about like "great butterflies." Mission accomplished! The young explorer was able to return to England with 35 cases of bulbs, rhizomes and tubers, a herbarium of 2,600 plants and seeds from over 300 species.

Veitch sent their new collector — now known as "Chinese" Wilson — on a new mission. This time he was to bring back a yellow poppy, *Meconopsis integrifolia* or lampshade poppy, from the Tibetan foothills. Wilson left in July 1903, traveling nearly 1,300 miles (21,000 km) in five and a half months, on raging rivers, and then through the foothills of the Himalayas. Suddenly, in the Ya-jia pass, buffeted by icy winds, he found the first yellow poppy, "the most gorgeous alpine plant extant." Ascending further, he found the high prairies were literally covered with the golden flowers.

Happy with this rapid success, the young plant hunter decided to look for another poppy, a red one. The *Meconopsis punicea* with its drooping petals had only been described by

(continued on page 148)

*A Tung tree (*Vernicia fordii *or* Aleurites fordii*) in flower on a Chinese hillside. Photograph by Wilson*

"At 11,000 feet I came across the first plant of Meconopsis integrifolia*! It was growing amongst scrub and was past flowering. I am not going to attempt to record the feelings which possessed me on first beholding the object of my quest to these wild regions ..."*

Ernest Henry Wilson

WILSON'S 50

Japan had many wonderful surprises for Ernest Wilson when he traveled there between 1911 and 1915. He found 63 species of cherry tree, but his finest discovery came in his search for the wild ancestors of azaleas, called *kurume*. On the tiny island of Kirishima, at the southern tip of the archipelago, he found an ancient garden, a museum, where there grew an unimaginable profusion of azaleas — salmon, white, pink, purple.... Amazed, Wilson selected 51 varieties, which, when introduced to the United States in 1920, were known as "Wilson's 50."

FACING PAGE

HERBARIUM PLATE

Davidia involucrata

This specimen of the dove tree, or handkerchief tree, was collected in western Szechwan, China, in 1908 by Wilson himself.

510

HERB. HORT. BOT. REG. KEW.

Davidia involucrata, Baill.

China: Western Szechuan, Coll. E.H. Wilson, 6.08.

bis.

one Russian traveler. The trip was terrible, admitted Wilson, "much harder than I had ever experienced." He fell ill, but still covered 550 miles (880 km) in 35 days and got his red poppy seeds. It was found to be extremely hard to cultivate, because the plant dies after it flowers, and its life cycle has ended. The golden poppy, however, has spread happily across Europe, especially in cool Scottish gardens.

On the trip home, on the banks of the Min River, Ernest Wilson had his first sight of the lily that would change his life. It was a tall bouquet of immaculate, majestic white flowers. "The blossoms of this lily transform a desolate and lonely region into a veritable flower garden," he gushed, and felt, rightly, that this discovery would "make his reputation." Sadly, Wilson's desire to save pennies led him to use inferior packaging and most of his 300 bulbs rotted on the voyage to England.

His next expeditions to China were devoted to finding conifers for Harvard University. Charles Sargent, the great botanist who was director of the prestigious school's Arboretum, sent him out with the freedom to collect "to increase knowledge" and provided him with the best modern photography equipment. Ernest Wilson's mission was to photograph landscapes, habitats and trees that were remarkable for their rarity or size. Imperial China was full of ginkgos, spruce, figs and centuries-old maples with altars where worshippers prayed.

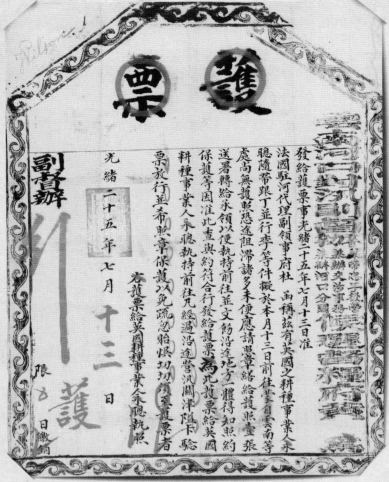

A pass issued in 1899 for travel to Yunnan

Wilson brought back many marvels from these years in Szechwan, including the magnificent *Rhododenron moupinense* (white with magenta edging) and other rhododendrons, the clematis *Montana* with its pink star shaped flowers, the Beauty Bush (*Kolkwitzia* , now *Linnea*), the first purple candelabra primula, the white magnolia (*wilsonii*), orchids, the first kiwi plants, etc. In all, he found some 1,200 species of trees and shrubs, 400 unknown species, four new genera and 1,600 herbarium specimens. China's botanical richness made it in Wilson's eyes "the mother of gardens," a "floral paradise" and the "kingdom of flowers."

In September 1910 he decided to deal with the regal lily once and for all, and set off for its only breeding ground, the Min valley. The lily was still there, still splendid, but the botanist was exhausted. On September 3, he wrote in his journal, "I am mightily fatigued by this life of wandering and I aspire to see the end of it. I have the feeling I have never done anything but run all over China." The next day a rock slide fell on his chair and the rocks fractured his right leg. Wilson crawled to shelter and made a temporary splint with a camera tripod. Then he was carried back to civilization by his porters, in a race against possible gangrene. They got him to the mission at Chengdu in 3 days, where he was cared for. Forever after he walked with what he poetically and philosophically called his "lily limp."

PHOTOREPORTER

The Director of the Arnold Arboretum insisted that Wilson take a large-format Sanderson field camera capable of recording both great detail and broad perspectives without distortion. It required a cumbersome wooden tripod and crates of heavy, fragile, glass-plate negatives. Between 1907 and 1922, Wilson took 2,488 photos in Southeast Asia. The more than 1,000 images from Imperial China showed trees, peasants, tea porters — an entire world that was soon swallowed up by the Communist revolution.

Tea porters. Photograph by Wilson

"*Above 11,000 feet, up to the summit of the mountains, 99 percent of the plants are rhododendrons. We saw thousands, hundreds of thousands of them. Their flowers were crimson, scarlet, flesh pink and silvery pink; some were yellow and others pure white.*"

Ernest Henry Wilson

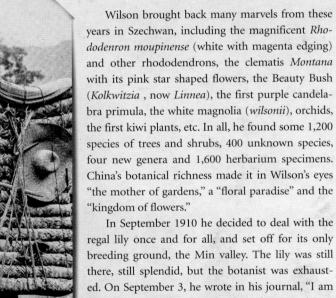

FACING PAGE

HERBARIUM PLATE

Meconopsis integrifolia

This poppy specimen was collected by Wilson himself during his 1907–09 expedition in China.

Meconopsis integrifolia (Maxim.) Franch.

Det. Kirstin B. Jork, 1997

Meconopsis integrifolia (Maxim.) Fr.

DET. G. Taylor 193 3

Wilson 3029.

No. 3029, ARNOLD ARBORETUM.
EXPEDITION TO CHINA, 1907-09.
Western Szechuan.

Meconopsis integrifolia Franch

1-3 ft. abundant

upland Pan-lan shan

W. of Kuan Hsien Alt. 11-13500ft

Coll. E. H. Wilson. 24/7/08

LOSING HIS HEAD OVER PRIMULAS

George Forrest in Yunnan

George Forrest was born in Falkirk, Scotland, in 1873 and was a young man who loved nature and the great outdoors. He apprenticed in a chemistry lab where he learned to identify, gather and dry botanical specimens. From 1891 to 1902 he was part of the Gold Rush in Australia, then returned to Scotland and began work at the Royal Botanic Garden in Edinburgh. In 1904 he was hired by an amateur collector of exotic plants and led his first expedition to Yunnan. He made seven more trips there in 28 years, escaping great danger and collecting over 30,000 species of plants. Forrest died from a heart attack in Yunnan in 1932, on his final voyage.

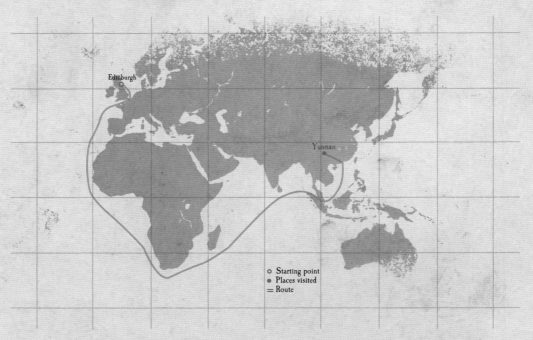

Edinburgh

Yunnan

○ Starting point
● Places visited
═ Route

George Forrest was short, robust, and, at 31, particularly tough. He had survived 10 years in the Australian Gold Rush and on returning to his native Scotland, he was seeking an outdoor job. He ended up accepting a job with the Royal Botanic Garden in Edinburgh, but leapt at the chance to collect exotic flowers for a rich Liverpool industrialist.

Forrest went to Yunnan, where the borders of China, Tibet and Burma met, to fulfill his client's orders. It was a remote, inhospitable region, marked by three steep-canyoned rivers, the Mekong, the Yangtze and the Salween, and mountains with 16,400 foot (5,000 m) peaks. The vegetation began at 9,850 feet (3,000 m): dwarf oaks, then an abundant alpine forest, then gigantic rhododendrons that were covered in snow when not in white flowers. Above them, there was nothing but grass, bare rock and glaciers.

In 1905 Yunnan was in the turmoil of war because the British had invaded Lhasa. Tibetan lamas tracked and killed strangers and anyone who cooperated with them. Forrest did not back down. He learned Chinese, gathered a group of collectors, and plunged into the forests. While his expedition was resting at a Catholic mission, the lamas attacked. Forrest, his team and two old priests tried to flee in the night, but they were caught, one by one, shot with poisoned arrows and decapitated. Miraculously, Forrest escaped, and for 8 nights he ran south, pursued by enormous Tibetan mastiffs. On the ninth night, he was taken in by Lissu villagers who disguised him and helped him

escape through "miles and miles of rhododendrons and primulas."

Although Forrest risked losing his head to the lamas' swords, he was able to harvest many of these plants. He collected more than 300 species of rhododendrons, including *Rhododendron griersonianum*, the ancestor of many of today's hybrids, and some 50 incredible primulas. These include *Primula vialii* with its bright pink spikes, *Primula bulleyana* with fluorescent orange flowers on a tall stem, and *Primula forrestii*, the one that bears his name and clings to the chalky cliffs of Yunnan.

In seven expeditions, assisted by squads of collectors, Forrest brought back many pounds of seeds and thousands of bulbs, some 31,000 plants in all, including 5,375 that were new to science. The list of his discoveries goes on and on: gentians, camellias, mahonias, asters, anemones, deutzias, iris, jasmine, conifers, etc. Europe's gardens were changed forever by George Forrest.

THE GARDEN OF EDEN OF RHODODENDRONS

The Yunnan region contains an unbelievable abundance of rhododendron species: giant, dwarf, climbing, and of every color. The further north Forrest went, the more he discovered. He eventually worked out a theory that there was one protected valley, somewhere to the north, that was the birthplace, the Garden of Eden, of the rhododendron genus.

IN THE LAND OF GIANTS

When Forrest found the giants, he named them *Rhododendron giganteum*. One of them was 79 feet (24 m) tall and had a girth of 7 feet (2 m). He felled one of these giants and its growth rings revealed its age: 280 years.

George Forrest at his camp

"The pines along the top of the ridges stand out as if limed by the hand of a Japanese artist. In the evening all the side slopes of the Mekong side are flooded with red and orange lights, which defy photography and would be the despair of a Turner."

George Forrest

FACING PAGE

HERBARIUM PLATE

Primula forrestii

This specimen was collected in western China by Forrest himself in 1910.

21 MAR 1915

7

YUNNAN, WEST CHINA   No. 5-5-6 3.
Coll. GEORGE FORREST
May. 1910.

Alt. 8.-10,3-00ft..
Locality Eastern flank of the Lichiang
Range. Lat-. 27° 15' N..

Primula Forrestii. Balf. fil.

ADVENTURES IN THE LAND OF THE BLUE POPPY

Frank Kingdon-Ward in the Himalayas

Frank Kingdon-Ward was born in 1885 in Manchester, where his father taught botany. Young Frank began his studies at Cambridge but was obliged to find work after his father died. He became a teacher in Shanghai, then joined an expedition on the Yangtze. In 1911, he went to collect plants in the Himalayas on behalf of the Bees Seed Co.. He explored China, Tibet, Burma and India for more than 40 years, sending occasional intelligence to the British government and writing 25 accounts of his travels. In 1955 he made his last climb up a mountain in Burma, to a height of 9,850 feet (3,000 m) and discovered more new species. He died three years later at the age of 73.

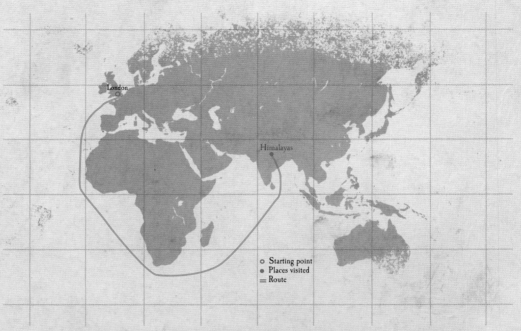

○ Starting point
● Places visited
═ Route

London

Himalayas

"There are places beyond the Brahmaputra where the white man has never set foot." Young Frank Kingdon-Ward heard these words from one of his father's friends, and they set the course for his future. In 1911, when he was 26, he was realizing his dream at the slow pace of mules, along the steep gorges of the Mekong River. The Tibetan rebellion continued and the Chinese authorities had equipped him with an escort of 3 "braves" armed with "a fan and a water pipe." The trails up to the high Himalayan plateaus led across torrents that had to be crossed on unstable bamboo bridges or terrifying rope bridges. But the trails were bordered with bamboo and mosses, ferns, orchids, lilacs, primulas, honeysuckles, barberry bushes "with tall pyramids of yellow flowers" and "deliciously scented white jasmine."

At one stop, the tenderfoot adventurer stepped away from his porters to chase a snow pheasant. A few turns later, he was lost. Panicked, he plunged deeper into the impenetrable jungle and wore himself out in thickets of giant bamboo. "Even the beautiful sight of masses of the blue *Primula sonchifolia* ... sometimes growing right in the icy water derived from the melting snow, failed to compensate me for this torture, or to rouse my enthusiasm." He spent the rainy night huddled under his tattered raincoat. The next day he was so hungry he sucked the nectar of rhododendron corollas, and ended up eating the flowers and leaves. Soon, he was twisted in painful cramps. Suffering from "extraordinary hallucinations," the botanist found himself "continually halting to step carefully over large

Mishmi porters crossing a stream in flood. Photography by Kingdon-Ward

boulders which did not exist except in [his] imagination, while in doing so [he] blundered clumsily into every obstacle" and "strange animals moved in the thickets." Finally, his luck led him to an isolated farm.

After sleeping heavily for 2 days and 2 nights, Kingdon-Ward was on the move again. He set out for the highlands covered with rhododendrons and potentillas, and on to the alpine prairies dotted with saxifrage and gentian. "Up here, at 17,000 feet, springing from amongst huge blocks of grey stone, I found the glorious Cambridge blue poppywort (*Meconopsis speciosa*), one of the most beautiful flowers in existence." From that first expedition, Frank Kingdon-Ward brought back 200 different species, of which 22 had not yet been seen by scientists. This coup launched his career as a collector and began more than 40 years of adventures during which he survived a fall from a cliff, having his tent crushed by a tree during a storm, being arrested by the Tibetans and experiencing a violent earthquake.

A LEGENDARY POPPY

The first blue poppy arrived in Geneva, sent by an English traveler who named it *Meconopsis napaulensis* (from Nepal). In 1866 Father Delavay found it again, after "weeks of walking toward the Tibetan plateau surrounded by 6,000 metre peaks. Whole days in a dense fog, losing his way on the edge of a precipice, then the burning sun that pains the eyes, after freezing nights." The sky-blue flower has four petals and, instead of drooping toward the earth, its stems stand up like halberds. It was officially named *Meconopsis betonicifolia*, and has kept that name even though Kingdon-Ward, thinking he had discovered it, called it *Meconopsis speciosa*.

"Sometimes the narrow path was enveloped in the shade of flowering shrubs and walnut trees, the branches breasting us as we rode, the air sweetened by the scent of roses which swept in cascades of yellow flowers over the summits of trees ..."

Frank Kingdon-Ward

FACING PAGE

HERBARIUM PLATE

Cassiope wardii
Marquand sp.

This specimen was collected in the eastern Himalayas by Kingdon-Ward himself.

HERB. KEW.

EASTERN HIMALAYA.
Coll. Capt. F. Kingdon Ward.
Presented by the Royal Botanic
ARDEN, Edinburgh, November, 1925.

Date 7 6/24

Altd. 14000-15000 feet

Locality Tewo La

Cassiope (sp. nov.) Wardii Marquand
On steep grassy and rock strewn alpine
slopes and ridges, exposed to sun and wind
but yet in flower.
20 6/24 Nyima La. In flower. Flowers cream,
with red calyx. Like no. 5663, but pedicels shorter

Field No 5752

F.K.W.

Cassiope Wardii Marquand sp. n.

Determinavit CWSN.

SAVING BIODIVERSITY, SEED BY SEED

Nikolai Vavilov on five continents

Nikolai Vavilov was born in Moscow in 1887, into a prosperous merchant family. He entered the Moscow Agricultural Institute in 1906, participated in the first expedition to the Caucasus, and introduced Darwin's theories to his country. Plant genetics was his passion, and he gained experience by working in the best English, German and French (Vilmorin) laboratories. He pursued his great project, collecting edible plants, from 1916 to 1940, and wrote about it in *Five Continents*. He founded and was director of the All-Union Research Institute of Applied Botany and New Crops, for which he was internationally famous. A victim of Stalin's purges, he was imprisoned and tortured and died in 1943.

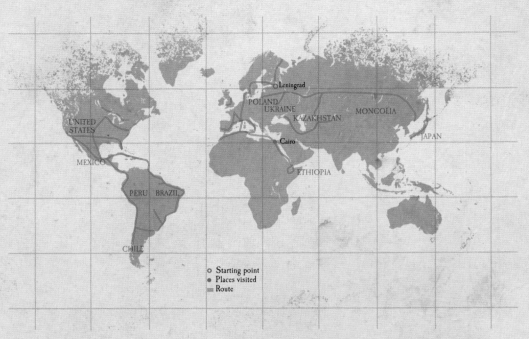

○ Starting point
● Places visited
═ Route

Russia was unstable in the early 20th century: the Russo-Japanese war, the 1905 revolution, World War I, the 1917 revolution, civil war... The people were in revolt but disorganized and repressed. Food shortages led to starvation. Nikolai Vavilov, the young agronomist passionate about genetics, felt an urgent need to develop agriculture, both by using new seeds and by saving existing varieties. Wheat, barley, rice, oats, potatoes, beets, cabbage, forage crops — all were available in the vast Russian Empire with its great variety of land and climate.

In 1916 Vavilov was sent to the Russian-Turkish front in the north of Iran and began collecting local cereal grains. In 1921, at 33, he was appointed Director of the Bureau of Applied Botany in Leningrad and undertook to make it a temple of science devoted to nourishing humanity. Seeking biodiversity, he sent his collectors around the world and himself explored 64 countries in two decades. In 1921, he went to the United States and Canada, in 1924 Afghanistan, between 1926 and 1929 he went around the Mediterranean (he was impressed by Spain), then Ethiopia and Yemen, China, Japan, Korea and Taiwan. He devoted 1930–33 to the Americas, from Manitoba to Uruguay. At the same time, his team was systematically exploring the USSR from the Black Sea to Siberia, from the Caucasus to the Arctic Circle. As a result, he enriched Russia's genetic heritage collection by 50,000 varieties of wild plants and 31,000 wheat

specimens, 10,000 varieties of corn (maize), 23,000 legumes, 18,000 vegetables, 12,000 fruits and 23,000 varieties of forage crops. Russia's agronomists and geneticists were the world's leaders.

But Stalin was at work and his brutal collectivization destroyed the peasants and their harvests. Those in power found a scapegoat in Vavilov. They accused him of undermining socialist reforms, arrested and tortured him, but he survived until January 1943. The staff of the research institute faithfully continued his work even as the German army besieged Leningrad. During that long siege, when 70,000 people died in the city, the scientists died of starvation while guarding cases and cases of grains and tubers, for the future good of humanity. Thanks to their sacrifice, the collections of the Vavilov Institute would give rise to more than 80 percent of the plant varieties cultivated in Russia today.

THE CRADLES THEORY

In 1935, Vavilov developed a revolutionary theory of the origin of plants. By studying plant varieties and the history of agriculture, he designated seven "centers of origin" for the world's plants. Southern Asia produced coconut, rice and sugar cane. From China came Chinese cabbage and soybeans; from India, cucumbers, eggplant and peas; from Iran, wheat, barley, oats and figs. The Mediterranean was the source of almonds; corn and tomatoes came from Central America and peppers and potatoes from South America.

Nikola Vavilov and his horse on an expedition across the USSR

"The intensive cultivation of eastern and southern Spain promoted selection of remarkable varieties, many of which are the best in the world ... onions, leguminous plants, olives ... which deserve the attention of Soviet plant breeders."

Nikolaï Vavilov

FLORA OF *Persia*. No.

Name: *White Cicer*

Native Name:

Locality: *Tabriz.*

Altitude:

Habit, Colour, Uses, etc.:

Cult. in Herb. Expt. Ground, Kew, 29.7.29.

Collector: *B. Gilliat-Smith.*　　Date:

THE LOST FLOWER OF THE DESERT
Théodore Monod in the Sahara

Théodore Monod was born in Rouen in 1902. His father and other family members were Protestant preachers. Théodore's intelligence was recognized early; he completed his studies and found a position at the Muséum national d'histoire naturelle, his "spiritual home," by age 20. While on a research expedition to the coast of Mauritania in 1923, he saw the Sahara for the first time. He never stopped exploring that desert, as a member of the camel corps and as a scientist, often alone and in extreme circumstances. He was the founder of the Institut français d'Afrique noire, an outstanding naturalist, an activist philosopher and a pacifist, and his immense body of work included 1,000 articles, 700 of them major scientific statements. He died in 2000 at the age of 98.

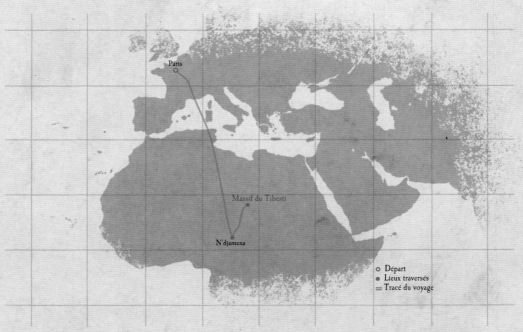

Paris

Massif du Tibesti

N'djamena

○ Départ
● Lieux traversés
═ Tracé du voyage

In March 1940, a young master corporal in the 3rd battalion of the *Tirailleurs sénégalaise* infantry was on patrol in the Tibesti Mountains, on the border with Libya, which was then in Italian hands. The dusty garments of a camel rider hid one of the most brilliant naturalists of the 20th century, a man with an encyclopedic memory and a mind always searching for discoveries. Although he was an expert in fish and crustaceans, Théodore Monod had fallen in love with the African desert, saying, "In the beginning, I was a zoologist, but as I traveled the dunes, I ended up collecting some of everything, including fossils and plants. Thus, I became a bit of a botanist, geologist, ethnologist and archaeologist."

Following the arid and winding wadi (dry riverbed) of Aïm Gongom, he made an unexpected discovery: a tiny stream, "which is rare in the Sahara," and a cliff covered with ferns and other plants. As always, he took specimens, pressed them between heavy paper and numbered them. Among them was one little flower on the end of flexible stems, which he took only "to be thorough."

Back at the Muséum, the little plant was found not only to be unknown, but from an unknown genus. It was named *Monodiella flexuosa* in honor of its discoverer. In later years its novelty was contested by other botanists who classified it among the *Centauria*. To settle the argument would require a DNA test, but this "debris" as Monod called it, was too small for a sample.

In November 1996, a 94-year-old man, nearly blind but with the impatience of a youth and the most enquiring of minds, climbed the Aïm Gongom gorge under a blazing desert sun. This time, he had not ridden on the back of a camel but in a 4x4, still a tiring journey "to the edge of inhabited world." While Libya had not been Italian for a long time, it was still hostile

to Chad, its neighbor. The border was a no-go zone and heavily mined. This clandestine expedition had only one purpose: finding the old naturalist's holy grail of botany, his little desert flower. Alas, there was persistent drought in the Tibesti range and the spring was only running drop by drop. Monod got out his magnifying glass and examined all the plants, in vain. He was cruelly disappointed but, as a scientist, he noted in his journal that "someone will find this species somewhere else, naturally. There are very few examples of a plant that exists in only place on the globe," and added, "but it is hard to imagine where it might be found." The search would go on, but in 2000 Théodore Monod made his "great departure, to the top of the mountain, above the clouds, into the light."

Théodore Monod in 1927 in the center of the Sahara

20,000 DISCOVERIES

Beginning with his first mission in Mauritania, Monod kept a numbered list of each plant, stone, engraving or shell he found, He did not want to die without having at least 20,000 specimens of flora and fauna. He succeeded: the last two specimens in his herbarium, collected in an oasis in Mauritania in 1998, were numbered 20272 and 20273.

BONESHAKER

When gathering plants on his long journeys by camel, Monod invented a press made of two boards strapped together. He wore it slung over his shoulder and called it his "Monod boneshaker system," or "Monod-shaker."

"I found a little plant that did not especially attract my attention: it is flexible and not very long; the flower is very small, and just to be thorough, I brought it anyway, although just the one specimen."

Théodore Monod

FACING PAGE

HERBARIUM PLATE

Bromus tibesticus Maire

This specimen was collected in Chad in 1940 by Monod himself.

in Bull. Soc. Hist. Nat. Afr. Nord, XXXIV. 140 (1943)

MPU - Maire   **Herbier Dr. René C. J. E. Maire (1878 – 1949)**   MPU - Maire
Herbier des plantes vasculaires d'Afrique du Nord
Université Montpellier II - Institut de Botanique - 163, rue A. Broussonet - F. 34090 MONTPELLIER
MPU   MPU

UNIVERSITÉ D'ALGER

HERBIER GÉNÉRAL

Bromus tibesticus Maire

Tibesti

TYPE

31-1-1940

Monod 7638   Dr R. Maire

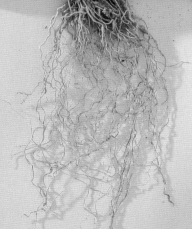

TCHAD
MPU

MPU   NOTULAE CRITICAE
Bromus tibesticus Maire
(= *Bromus pectinatus Thunb.*)
DATUS : 24.07.2007   AUCTOR : A. Gely
Herbiers Institut de Botanique Université Montpellier 2 - 163, rue Auguste Broussonet F-34090 MONTPELLIER MPU

THE AMAZON MOONFLOWER
Margaret Mee in Brazil

Margaret Ursula Mee was born near London in 1909. Her natural artistic talent was set aside while she worked for the British Communist Party. From the 1930s until the beginning of World War II, she supported the unemployed and the Spanish Republicans in their struggle against fascism. Later, she returned to painting in order to earn a living and studied at a well-known art school. In the 1950s she moved to Brazil where, fascinated by tropical flora, she led 15 expeditions into the untouched forest and acquired an international reputation as a botanical painter. Hers was a vibrant voice in favor of saving the Amazon until she died in a car accident in 1988.

○ Starting point
● Places visited
═ Route

A little motorboat traveled up the black waters of a tributary of the Amazon one day in May 1988, carrying a fragile-looking English lady of 78, wearing a flowered skirt. Her silver hair was tied back with a green velvet cord and her blue eyes searched the tropical forest that she knew like the back of her hand. Somewhere along the 1,300 miles (2,080 km) of the Rio Negro "her" cactus was blooming. The *Selenicereus wittii*, a moonlight cactus, owes its name to Selena, the Greek goddess of the full moon. Its flowers open for only one night. In the 36 years Margaret Mee had been traveling through the Amazon in a canoe, she had only seen the plant three times, always too early or too late to see it in flower.

Finally, they found the rare epiphytic cactus clinging to the bark of a tree, its purple leaves bordered with thorns and its long tubular flower stems drooping toward the river. A folding chair was set up on the roof of the boat and Margaret Mee sat there, drawing pad on her knees, pencils and paintbrush in her hands. By the light of a flashlight, she finally witnessed the miracle, saying, "The flower exhaled a powerful fragrance and we were all struck motionless by its beauty, delicacy and surprising size." At midnight, the enormous white flower opened up. At 3 in the morning, it was already wilted. Until morning, the painter used her consummate talents and absolute vigilance to create the portrait of the moonflower. Scientifically correct, as well as luminous and inspired, it captures the ephemeral soul of the plant.

That night was the highlight of an exceptional career as an artist, botanist and explorer. Margaret Mee had made 14 expeditions into the Amazon basin, and even to the Neblina peak on the Venezuelan border, often traveling for months with only her Indian guide, her paint box and her straw hat. During these solitary journeys she discovered unknown plants: the *Neoregelia margaretae* crimson jewel, and *Aechmea meeana*, which are named for her, as well as *Aechmea polyantha*, a large, amphora-shaped bromeliad, and *Heliconia chartacea,* with indigo blue flowers. Shipwrecks, malaria, hepatitis and scorpion stings could not dissuade her from hours spent studying an orchid petal, a bromeliad leaf or the cascading stem of a heliconia. After one encounter with gold prospectors, she simply added a little 1.25 inch (32 mm) pistol to her backpack. The dangers of the Amazon could not shake the calm courage of Margaret Mee, but only 6 months after her moonflower night, she died in a car accident in England.

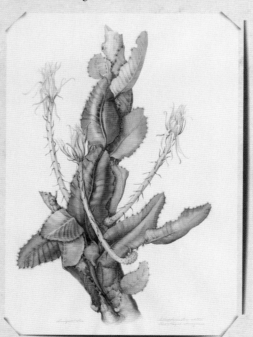

Selenicereus wittii *painted by Margaret Mee*

SAVING THE AMAZON

In the 1970s, Margaret Mee was aware she was painting a threatened world. The natural rainforest was "being attacked by axe and fire" so badly that some species were about to disappear. "Will the exquisite beauty of this species save it from extinction?" she wrote about the magnificent *Gustavia pulchra* in 1979. In 1984, on the way to Manaus, she lamented, "Where the virgin forest used to stand, there is only a blackened lake full of huge skeletons...." Revolted by the "devastation that Man inflicts everywhere he sets foot," the artist joined the crusade against multinationals, banks and unscrupulous developers.

"The surface of the water was like a mirror and it was nearly impossible to distinguish the real from the reflection. Orchids and bromeliads were everywhere and among them I was lucky enough to discover Catasetum punctatum. I succeeded in painting these flowers before night fell, and after that, we dined on piranhas."

Margaret Mee

FACING PAGE

HERBARIUM PLATE

Selenicereus wittii, formerly *Strophocactus wittii*

This specimen was collected in Brazil and mounted in 1964.

University of California Botanical Garden, Berkeley
Accession Number 51.950-1

Strophocactus wittii Br. & R.

Determination by Paul C. Hutchison
Date prepared Dec. 4, 1964   by D. M. Hutt
Source: New York Botanical Garden. Plants
received and grown under glass at UCBG.

_____

Collector:
Field collection data: Brasil.

Cult. Notes; Fragrance light and sweetly
fetid, "moth smell". Nectar in beads in lower
4 inches of tube.

57

THE MAN WHO (RE)PLANTED TREES

Tony Kirkham in East Asia

Anthony Stuart Kirkham was born in Lancashire in 1957. At the age of 11 he fell in love with the chestnut tree at his school, which furnished "a great deal of pleasure, information and chestnuts to play with." He was also fond of sculpture but preferred standing wood, and studied forestry and arboriculture. In 1981 he obtained the Kew diploma and began his career at the Arboretum. He took responsibility for 14,000 living specimens, including some very old trees that were difficult to maintain. Between long-distance voyages — to Chile, South Korea, Taiwan, Japan, Russia, China, Turkey, Australia and the United States, among others — Tony Kirkham has shared his great experience of trees with the general public, on the BBC as well as on the grounds at Kew.

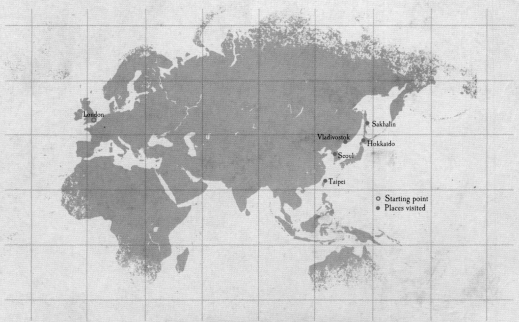

London
Sakhalin
Vladivostok
Hokkaido
Seoul
Taipei

○ Starting point
● Places visited

Tony Kirkham (in the tree) and Mark Flanagan (below) collect the fruit of a Sorbus sp. (mountain ash or rowan) in Szechwan, China, 2001.

The most destructive storm England had seen in 250 years began at 2 a.m. on October 16, 1987. In less than 5 hours it had swept the south of the country, leaving behind ruin and desolation. The royal gardens at Kew and their satellite at Wakehurst were struck a terrible blow. In the morning, the broken-hearted gardeners counted their dead. At Kew, 800 trees were downed, uprooted or broken, and 1,500 more at Wakehurst. And what trees they were! All were exceptional specimens and botanical treasures, and many were historic monuments.

After the sadness came the time for rebuilding. Unexpectedly, the inventory of losses also revealed gaps in the collections. The idea began to grow of going back to the garden's century-old tradition and sending new expeditions to far away lands. Two years after the Great Storm, one of the arboretum's managers, Tony Kirkham, along with his colleague, the late Mark Flanagan, took off for South Korea. Their mission was to collect herbarium specimens and seeds that would eventually become trees. Working from a list of priorities, they tracked their "targets." Modern collectors are helped enormously by airplanes, 4x4s and GPS, but they still suffer from cold, rain, fog, sunshine, lost guides, landslides, flooded rivers and persistent mosquitoes. But the harvest exceeded all expectations and produced seed from maple, beech, magnolia, arborvitae, mountain ash (rowan), rhododendron, mock orange, deutzia, viburnum, dogwood and even a cotoneaster discovered by the illustrious Ernest Wilson.

With that success behind him, Tony Kirkham went to Taiwan in 1992. That island is home to the rare Taiwanese birch and an abundance of endemic conifers, including yew, fir and cypress. That autumn was very fruitful for tree seeds; Kirkham put on his climbing harness to fetch cones from the tips of immense prickly trees. In 1994 it was the forests of eastern Siberia and Sakhalin Island that gave him their pine, oak, maple, alder, birch and mountain ash. Finally, in 1997, Tony Kirkham explored Hokkaido, Japan's large northern island, where he collected cedar, sumac, and new and rare varieties of maple and birch. More than 1,220 trees — 426 species — from vigorous wild sources were replanted at Kew and Wakehurst thanks to these four voyages. As the third millennium dawned, the hearts of these venerable arboretums beat with renewed energy.

THE ICE AGE

Why does one single mountain in Szechwan contain more species of trees than all of Great Britain? The answer comes from the Pleistocene era and the successive glaciers that wiped out much of the European flora. Hemmed in by the Alps and Pyrenees, backed up against the Mediterranean, the endemic trees, flowers and fungi could not retreat any further when the glaciers came down from the north. They died before better weather returned when they might have repopulated their territory, as the plants of Asia and America did.

"We climbed to 620 meters, above a charming, colorful Buddhist temple. I was enchanted to find Lindera obtusiloba *wearing its butter-yellow autumn foliage, and my favorite birch,* Betula dahurica, *its wonderful pale, scaly bark lit up by the noonday sun coming through its leaves.*"

Tony Kirkham

FACING PAGE

HERBARIUM PLATE

Cephalotaxus harringtonii, formerly *Cephalotaxus wilsoniana* Hayata

This specimen of Japanese pine was collected by Kirkham and Flanagan in Taiwan in 1992.

BR 21740

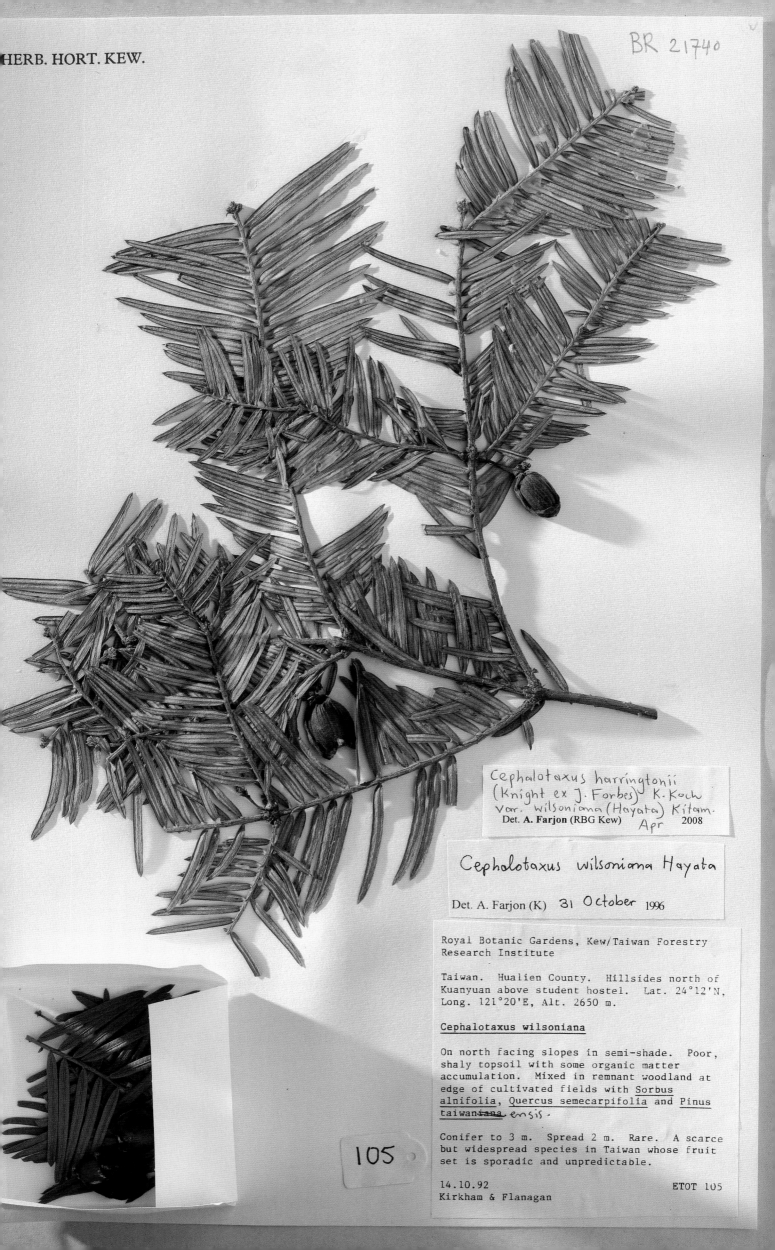

Cephalotaxus harringtonii
(Knight ex J. Forbes) K. Koch
Var. *wilsoniana* (Hayata) Kitam.
Det. A. Farjon (RBG Kew)
Apr 2008

Cephalotaxus wilsoniana Hayata

Det. A. Farjon (K) 31 October 1996

Royal Botanic Gardens, Kew/Taiwan Forestry
Research Institute

Taiwan. Hualien County. Hillsides north of
Kuanyuan above student hostel. Lat. 24°12'N,
Long. 121°20'E, Alt. 2650 m.

Cephalotaxus wilsoniana

On north facing slopes in semi-shade. Poor,
shaly topsoil with some organic matter
accumulation. Mixed in remnant woodland at
edge of cultivated fields with Sorbus
alnifolia, Quercus semecarpifolia and Pinus
taiwaniana ensis.

Conifer to 3 m. Spread 2 m. Rare. A scarce
but widespread species in Taiwan whose fruit
set is sporadic and unpredictable.

14.10.92   ETOT 105
Kirkham & Flanagan

105

GLIDING ABOVE THE CANOPY

Francis Hallé on Madagascar

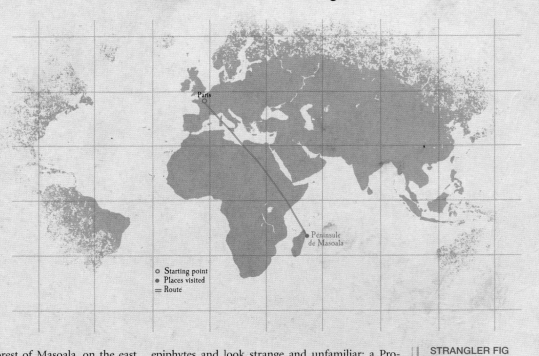

Francis Hallé was born near Paris in 1938 and was an instructor at Glénans sailing school before studying biology at the Sorbonne and at the University of Abidjan. He became a professor of botany at Montpellier and as a dendrologist (who studies woody plants) he has concentrated on tropical rainforests, the old-growth forests in particular. From 1996 to 2011, Francis Hallé led scientific missions to the Radeau des cimes (treetop raft) above the canopy of the tropical forest in French Guiana, Gabon, Madagascar, Panama and Laos. He founded the Wild Touch Association and has stated, "When we love something, we protect it better." He has written many books including *Éloge de la plante (In praise of plants)* and *Plaidoyer pour l'arbre (In defence of trees)* and was working on a film about threatened rainforests.

Paris

Péninsule de Masoala

○ Starting point
● Places visited
═ Route

As dusk fell, the forest of Masoala, on the east coast of Madagascar, resonated with the "admirable concert of thousands of frogs, night birds, monkeys and insects." At the top of the tall trees, scientists were preparing to savor an exceptional, though rather humid night in November 2010. They were in a strange nest that consisted of a 6,460 square foot (600 sq. m) polypropylene net surrounded by a huge tube inflated with compressed air. This treetop raft extended over the canopy, which looked like a field of leafy sheep. "The Milky Way was above us, and the fireflies below; it was true happiness," exulted the botanist, Francis Hallé.

This new way to navigate on treetops was the result of an encounter with Dany Cleyet-Marrel, an exceptional pilot "who's not afraid of trees" and with Gilles Ebersolt, an "astonishing" architect and inventor. Together, the three designed devices that would give them access to this still-unexplored area of the world: the forest canopy. A rainbow-colored compressed-air dirigible 165 feet (50 m) long was the moving force, and it pulled the "sled." This smaller vessel was suspended from the dirigible raft and was towed over the surface of the canopy where the scientists could harvest at will. The remaining parts of the arrangement were a "treetop bubble," a helium balloon that could take a scientist from tree to tree by sliding along a rope, and the passenger compartments made of metal tubes that function as office, kitchenette, observatory and cocktail-hour patio.

Early in the morning, the raft floats in the mist, but the sun strikes the treetops as soon as it rises. The intense light and high humidity makes the canopy the most lively place in the world, infinitely richer than the oceans. "The old-growth forest of Masoala shows us its very best: the trees are covered with mosses and epiphytes and look strange and unfamiliar: a Proteaceae *Dilobeia* with dichotomous leaves; a Clusiaceae *Mammea* that is monocaulous and cauliflorous; a twining bamboo; an epiphytic Araliaceae *Schefflera* never before seen on Madagascar; a microphyllous Sapotaceae *Capurodendron*," and the "incredible flowers of the Symphonia." In their floating laboratory, the botanists prepared the specimens they harvested, and drew and photographed them for later study on the ground. Other researchers discovered a new species of frog, captured a rare lemur and tree ants, and found new molecules for perfumes and drugs. The mission was a success even though the canopy did not deliver all its treasures in 6 weeks.

The dirigible moves the sled, enabling the scientists riding on it to gather specimens.

STRANGLER FIG

In tropical forests like those in French Guiana, certain epiphytic plans like the strangler fig grow in bright sunlight on the treetops, then send their roots toward the soil, 100 to 130 feet (30 to 40 m) below, to seek water and minerals. They eventually kill their host and take its place.

PRIMARY AND ELEMENTAL

Francis Hallé says the disappearance of old-growth tropical forests, those that have never been exploited by humans, is a disaster. A secondary forest would have to grow for 700 years to match the original forest. In French Guiana, large swaths are being deforested and Hallé says France must bear the responsibility.

FACING PAGE

HERBARIUM PLATE

Symphonia globulifera

This specimen comes from the canopy at Paracou, French Guiana. It was collected by Hallé himself during the mission from October to December 1996 using the sled system.

"What an astonishing memory: living 28 metres above the ground in a comfortable capsule among the branches of a great tree. We had a splendid view of the forest, the ocean and the dirigible as it took off. The branches were covered with orchids and a big blue bird came to see us."

Francis Hallé

150 *Symph* *...bulifera*.   CLUSIACEAE

boutons floraux rouges.

...Pascou, Luge du 14.XI.96,1996

n°6 SIG, arbre de ~ 32m de hauteur

D3

BIBLIOGRAPHY

GENERAL REFERENCES

Collective, *L'Herbier du monde, Cinq siècles d'aventures et de passions botaniques au Muséum d'histoire naturelle*, L'Iconoclaste, Les Éditions du Museum, 2004

Collective, *Passions botaniques*, Éditions Ouest-France, 2008

Y.-M. Allain, *Voyages et survie des plantes au temps de la voile*, éditions Champflour, 2000

Burton, Cavendish, Stonehouse, *Les Grands Explorateurs*, PML éditions, 1995

Rosemary Burton, Richard Cavendish, Bernard Stonehouse, *Journeys of the Great Explorers*, Facts on File, 1992

Lucienne Deschamps, *Botanistes voyageurs*, Aubanel, 2008

Carolyn Fry, *The Plant Hunters: The Adventures of the World's Greatest Botanical Explorers*, University of Chicago, 2013

Carolyn Fry, *The Plant Hunters: The Adventures of the World's Greatest Botanical Explorers*, University of Chicago, 2013

Joëlle Magnin-Gonze, *Histoire de la botanique*, Delachaux et Nieslé, 2004

Jean-Marie Pelt, *La Cannelle et le Panda*, Fayard, 1999.

Anthony Rice, *Voyages of Discovery: Three Centuries of Natural History Exploration*, Clarkson Potter, 1999

Catherine Vadon, *Aventures botaniques d'outre-mer aux terres atlantiques*, Jean-Pierre Gyss éditeur, 2002

WORKS *by or about* EACH EXPLORER

Adanson, Michel

Adanson Michel, *Histoire naturelle du Sénégal : coquillages : avec la relation abrégée d'un voyage fait en ce pays pendant les années 1749, 50, 51, 52 et 53*

Armange, Mathurin

Catherine Vadon, *Aventures botaniques d'outre-mer aux terres atlantiques*, Jean-Pierre Gyss éditeur, 2002

Banks, Joseph and Solander, Daniel

Journal d'un voyage autour du monde, fait par MM. Banks & Solander, Anglois, en 1768, 1769, 1770, 1771

Relation des voyages entrepris par ordre de Sa Majesté britannique... pour faire des découvertes dans l'hémisphère méridional, et successivement exécutés par le commodore Byron, le capitaine Carteret, le capitaine Wallis & le capitaine Cook, dans les vaisseaux le Dauphin, le Swallow & l'Endeavour : rédigée d'après les journaux tenus par les différens commandans & les papiers de M. Banks

Belon, Pierre

Travels in the Levant: The Observations of Pierre Belon of Le Mans on Many Singularities and Memorable Things Found in Greece, Turkey, Judaea by Pierre Belon (Author), Alexandra Merle (Introduction), James Hogarth (Translator), Hardinge Simpole Limited, 2012

Belon Pierre, *Les observations de plusieurs singularitez et choses mémorables, trouvées en Grèce, Asie, Judée, Égypte, Arabie et autres pays étranges*

Voyage au Levant : les observations de Pierre Belon du Mans, texte établi et présenté par Alexandra Merle, éditions Chandeigne, 2001

Yvelise Bernard, *L'Orient du XVIᵉ siècle à travers les récits des voyageurs français...*, Éditions de L'Harmattan, 1988

De La Billardière, Jacques

Relation du voyage à la recherche de La Pérouse, fait par ordre de l'Assemblée constituante pendant les années 1791, 1792 et pendant la 1re et la 2e année de la République françoise, par le citoyen Labillardière
Two volumes of a very nice atlas.

Bligh, William

Yves Kirchner, *L'Histoire vraie des mutins de la Bounty*, Découvertes Gallimard

William Bligh, *A narrative of the mutiny on board his majesty's ship Bounty, by William Bligh*

William Bligh, *A voyage to the South Sea undertaken by command of His Majesty, for the purpose of conveying the bread-fruit tree to the West Indies, in His Majesty's ship the Bounty, commanded by Lieutenant William Bligh, 1792* https://archive.org/stream/voyagetosouthsea00blig#page/6/mode/2up

de Busbecq, Augier

The Turkish Letters of Ogier Ghiselin de Busbecq: Imperial Ambassador at Constantinople, 1554-1562

Letters of Ogier Ghislain de Busbecq to the Holy Roman Emperor Maximilian II

Anna Pavord, *La Tulipe*, Actes Sud 2001

Cailliaud, Frédéric

Voyage à l'Oasis de Thèbes et dans les déserts situés à l'Orient et à l'Occident de la Thébaïde fait pendant les années 1815, 1816, 1817 et 1818, par M. Frédéric Cailliaud (de Nantes)

Voyage à Méroé, au Fleuve Blanc au-delà de Fazoql, dans le midi du royaume de Sennâr, à Syouah, et dans cinq autres oasis : fait dans les années 1819, 1820, 1821 et 1822.

De Candolle, Augustin

Le Voyage de Tarbes, 1807, Première grande traversée des Pyrénées, en voyage dans le Midi de la France, retranscrit par Alain Bourneton, Loubatières, 1999

Christopher Columbus

The letters of Christopher Colombus on the discovery of the new world.

Dampier, William

A voyage to New Holland : the English voyage of discovery to the South Seas in 1697

A new voyage around the world

Darwin, Charles

Patrick Tort, *Darwin et la science de l'évolution*, Découvertes Gallimard, 2000.

Voyage d'un naturaliste autour du monde : Fait à bord du navire le Beagle de 1831 à 1836, éditions La Découverte, 2003.

The voyage of the Beagle in The Complete Work of Charles Darwin Online (http://darwin-online.org.uk/life20.html)

John Van Wyhe, *Darwin*, Éditions Guy Trédaniel, 2009

David, Armand

Journal de mon troisième voyage d'exploration dans l'Empire chinois, par M. l'abbé Armand David

Dumont d'Urville, James and Lesson, René Primavère

Voyage de découvertes de l'Astrolabe exécuté par ordre du Roi, pendant les années 1826-1827-1828-1829, sous le commandement de M.J. Dumont d'Urville

Fairchild, David

The World was my garden. Travels of a plant Explorer, Scribners, 1939

The World Grows Round My Door, Scribners, 1947

Exploring for Plants, Macmillan, 1930

Forster, JG and JR

Voyage dans l'hémisphère austral et autour du monde (1778)

A Voyage round the World in His Britannic Majesty's Sloop Resolution, Commanded by Capt. James Cook, during the Years, 1772, 3, 4 and 5

Fortune, Robert

La Route du thé et des fleurs, Petite bibliothèque Payot, 1994

Le Vagabond des fleurs, Trois années dans la Chine du thé, de l'opium et des fleurs, Petite bibliothèque Payot, 2003

Grant, James Augustus

A walk across Africa, or domestic scenes from my Nile journal

The botany of the Speke and Grant expedition

Hallé, Francis

Journal de Francis Hallé, (http://www.radeau-des-cimes.org)

Francis Hallé, avec Dany Cleyet-Marrel et Gilles Ebersolt, *Le Radeau des cimes. L'exploration des canopées forestières*, Lattès, 2000

Francis Hallé et David Dellas, *Arbres et arbustes en campagne*, Actes Sud, 2010

Francis Hallé, *Plaidoyer pour l'arbre*, Actes Sud, 2005

Hatshepsut

Caroline Fry, *Chasseurs de plantes*, Prisma presse, 2010.

English edition: André Deutsch - Kew Garden 2009

Nathalie Beaux, *Le Cabinet de curiosités de Thoutmosis III : plantes et animaux du "Jardin botanique de Karnak"*, Leuven, Peeters, 1990

Humboldt, Friedrich Wilhelm Heinrich Alexander & Bonpland, Aimé Jacques Alexandre Goujaud

Voyage aux régions équinoxiales du nouveau continent, fait en 1799, 1800, 1801, 1802, 1803 et 1804, par A. de Humboldt et A. Bonpland

Nicolas Hossard, *Alexander von Humboldt et Aimé Bonpland — Correspondance 1805–1858*, Éditions L'Harmattan, Paris, 2004

De Jussieu, Joseph and de La Condamine, Charles Marie

Journal du voyage fait par ordre du roi, à l'équateur, by Charles Marie de La Condamine

Florence Trystram, *Le Procès des étoiles*, Petite bibliothèque Payot, 2001

Kingdon-Ward, Frank

Frank Kingdon-Ward, *The Land Of The Blue Poppy, Travels of a naturalist in Eastern Tibet*, University Press of Cambrige, 1913

Kirkham, Anthony

Mark Flanagan & Tony Kirkham, *Plants from the edge of the world*, Timber Press, 2005

Mark Flanagan & Tony Kirkham, *Wilson's China a century on*, Kew publishing, 2009

La Pérouse, Jean-François

Voyage de La Pérouse autour du monde

Livingstone, David

David Livingstone, *A Popular Account of Dr. Livingstone's Expedition to the Zambesi and Its Tributaries And of the Discovery of the Lakes Shirwa and Nyassa (1858-1864)*

The letters and papers of Livingston can be found at http://www.livingstoneonline.ucl.ac.uk/

Henry M. Stanley, *How I Found Livingstone: Travels Adventures and Discoveres in Central Africa*

De Lobel, Loicq

Loicq de Lobel, *Le Klondyke, l'Alaska, le Yukon et les îles Aléoutiennes*, Bulletin de la Société de géographie (Paris) 1899, tome 20, série 7

Loicq de Lobel, "L'hiver au Klondike", in *Le Monde illustré*, n° 2293, March 1901

Pierre-Christian Guillolard, "L'herbier de la ruée vers l'or"

in *L'Herbier du monde, Cinq siècles d'aventures et de passions botaniques au Muséum d'histoire naturelle*, L'Iconoclaste, Les éditions du Museum, 2004

Low, Hugh

Nigel Barley, *White Rajah*

Steven Runciman, *The White Rajahs: A History of Sarawak from 1841 to 1946*, 2011

Magellan, Ferdinand

Antonio Pigafetta. *The Philippine Islands, 1493-1898, Volume 33, 1519-1522 / Explorations by early navigators, descriptions of the islands and their peoples, their history and records of the catholic missions, as related in contemporaneous books and manuscripts, showing the political, economic, commercial and religious conditions of those islands from their earliest relations with European nations to the close of the nineteenth century* https://archive.org/details/philippineislan96bourgoog

Masson, Francis

Francis Masson, "An Account of Three Journeys from the Cape Town into the Southern Parts of Africa", article in the journal, *Philosophical Transactions of the Royal Society of London*, 1776: http://rstl.royalsocietypublishing.org/content/66/268.full.pdf

Michaux, André

Journal of André Michaux, 1793–1796, published 1904

Mee Margaret

Margaret Mee, I*n Search Of Flowers of the Amazon Forest*, Nonesuch Expeditions Publisher: Intl Specialized Book Service Inc; Reprint edition (November 1987)

Ruth Stiff and Laura Ponsonby, *Margaret Mee and Marianne North: their intrepid explorations and paintings*, Royal Botanic Gardens, Kew, 2007

Ruth Stiff, *The Flowering Amazon — Margaret Mee Paintings from the Royal Botanic Gardens, Kew*, Kew publishing, 2004

Monod, Théodore

Maximilien Dauber, *Le Vieil Homme et la petite fleur. Théodore Monod, sa dernière grande aventure*, éditions Nevicata, 2011

Roland Billard and Isabelle Jarry, *Hommage à Théodore Monod, naturaliste d'exception*, in coordination with, éditions du Muséum national d'histoire naturelle, 1997

Jean-Claude Hureau, *Le Siècle de Théodore Monod*, éditions MNHN/Actes Sud 2002

North, Marianne

Ruth Stiff and Laura Ponsonby, *Margaret Mee and Marianne North: their intrepid explorations and paintings*, Royal Botanic Gardens, Kew, 2007

Michelle Payne, *Marianne North, a very intrepid painter*, Kew Publishing, 2011

Plumier, Charles

Charles Plumier, *Description des plantes de l'Amérique, avec leurs figures*
Par le père J.-B. Labat Nouveau Voyage aux Isles Françoises de l'Amérique

Polo, Marco

The Travels of Marco Polo the Venetian, Everyman, 1908

Von Siebold, Philipp

Manners and customs of the Japanese, in the nineteenth century; from recent Dutch visitors of Japan, and the German of Dr. P. F. von Siebold. [Edited by Mrs. W. Busk.], Publisher: British Library, Historical Print Editions, 2011

Speke, John

The Journal of the Discovery of the Source of the Nile , 1864 https://archive.org/stream/journal00ofdiscovespekrich#page/n5/mode/2up

Spruce, Richard

Seaward & Fitzgerald, *Richard Spruce (1817-1893): Botanist and Explorer*, Royal Botanic Gardens Kew, 1996

Notes of a botanist on the Amazon & Andes by Richard Spruce, edited by Alfred Russel Wallace

Report on the expedition to procure seeds and plants of the Cinchona succirubra, or red bark tree

Theophrastus

Theophrastus, Enquiry into Plants, Volume I: Books 1–5, Loeb Classical Library 70, 1916

Thorel, Clovis

Francis Garnier, *Voyage d'exploration en Indo-Chine: effectué par une commission française présidée par le capitaine de frégate Doudart de Lagrée*

Clovis Thorel, *Notes médicales du voyage d'exploration du Mékong et de Cochinchine*

Thunberg, Carl

Japan Extolled and Decried: Carl Peter Thunberg and the Shogun's Realm, 1775–1796 by Carl Peter Thunberg (Author), Timon Screech (Editor), Routledge, 2005

Tournefort, Joseph

A Voyage into the Levant... (1741 translation) https://archive.org/details/avoyageintoleva00tourgoog

Tradescant, John

The John Tradescants, gardeners of the Rose and Lily Queen, Peter Owen Publishers, 1984, 2006

Vavilov, Nikolaï

Gary Paul Nabhan, *Where Our Food Comes From : Retracing Nikolay Vavilov's Quest to End Famine*, Island Press, 2008

Vavilov, Nikolai, *Five Continents*, (1996 translation)

Wickham, Henry

Joe Jackson, *The Thief at the End of the World: Rubber, Power and the Seeds of Empire*, Viking/Penguin, 2008

Wilson, Ernest

Mark Flanagan & Tony Kirkham, *Wilson's China a century on*, Kew publishing, 2009

Wilson Ernest Henry, *Plant Hunting*, 1927

Wilson Ernest Henry, *China, mother of gardens*, 1929

Wilson Ernest Henry, *Naturalist in western China, with vasculum, camera, and gun; being some account of eleven years' travel, exploration, and observation in the more remote parts of the Flowery kingdom 1913*

Many of these works are available online at: http://gallica.bnf.fr ; http://books.google.fr ; http://openlibrary.org ; http://www.archive.org ; http://www.gutenberg.org

PHOTOGRAPHIC CREDITS

All photographs of herbarium pages ©Yannick Fourié

Photographs of botanical explorers:

HATSHEPSUT
For both images (bust of Hatshepsut and relief at Deir-el-Bahari) ©Université libre de Bruxelles — Iconothèque numérique — Collection Égypte. Photo Rolland Tefnin

FERNANDEZ DE OVIEDO
Portrait : Fortaleza Ozama - Museo de Armas de Santo Domingo @ Ciudad Colonial Santo Domingo, República Dominicana

CHARLES PLUMIER
Portrait : ©Bibliothèque de l'Académie de médecine.

JOSEPH DE TOURNEFORT
Main image: ©Bibliothèque centrale M.N.H.N. Paris

JOSEPH DE JUSSIEU
Main image: ©Bibliothèque centrale M.N.H.N. Paris

PIERRE NOËL D'INCARVILLE
Main image: ©Bibliothèque centrale M.N.H.N. Paris

PIERRE POIVRE
Portrait: ©BIU Santé (Paris)

PHILIBERT COMMERSON
Portrait: ©BNF

JOSEPH BANKS AND DANIEL SOLANDER
Image at top of page 64: ©The Natural History Museum, London

FRANCIS MASSON
Portrait: © With permission of the Linnean Society of London.

JEAN-FRANÇOIS DE LA PÉROUSE
Main image: © Photo Josse/ Leemage

WILLIAM BLIGH
Painting of the mutineers setting Lt Bligh adrift: ©National Maritime Museum, Greenwich/ Leemage

ALIRE RAFFENEAU-DELILE
Portrait: ©Bibliothèque centrale M.N.H.N. Paris

MERIWETHER LEWIS
Main image: ©American Philosophical Society

FREDERIC CAILLIAUD
Main image: ©Photo Josse/ Leemage

JULES DUMONT D'URVILLE
Main image: ©Costa/ Leemage

CHARLES DARWIN
Drawing of the tortoise: ©The Natural History Museum, London

MATHURIN ARMANGE
Main image: ©Archives municipales de Nantes – 01 – 16144

DAVID LIVINGSTONE AND JOHN KIRK
Main image, page 122: ©Trustees of the National Library of Scotland

CLOVIS THOREL
Portrait: ©Société de géographie – SG Portrait – 1573

ARMAND DAVID
Photograph of Armand David in Chinese dress: © Société de géographie – SG Portrait – 1670

DAVID FAIRCHILD
Photographs from the book *The world was my garden; travels of a plant explorer,* by David Fairchild.

GEORGE FORREST
Portrait and main image: ©Royal Botanic Garden Edinburgh

FRANK KINGDON-WARD
Portrait and main image: ©Royal Geographical Society

NIKOLAÏ VAVILOV
Main image: from the book *Five continents* by Nicolaï Vavilov, with permission of the N.I. Vavilov Institute of Plant Industry.

THÉODORE MONOD
Portrait and main image: photographs ©JeanMarc Durou. The publisher was unable to contact the photographer.

MARGARET URSULA MEE
Portrait: ©South American Pictures/Tony Morrison/ MMXX0467, Margaret Mee on Rio Negro, Amazon 1988.

ANTHONY STUART KIRKHAM
Portrait and main image: ©Tony Kirkham

FRANCIS HALLÉ
Portrait and main image: ©Laurent Pyot

IN THE INTRODUCTION: VOYAGES TO PLANTS UNKNOWN

Page 10, 1875 map: ©BNF/Société de géographie
Page 17, photograph of seeds: ©M.N.H.N. Patrick Lafaite

ALL OTHER IMAGES ARE FROM PERSONAL COLLECTIONS OR FROM THE LIBRARY OF THE ROYAL BOTANIC GARDENS, KEW, LONDON

INDEX